Royal Horticultural Society

THE wildlife TRUSTS

BIRDS
IN YOUR
GARDEN

THINK
BOOKS

THINK
BOOKS

First published 2007 by Think Books
an imprint of Pan Macmillan Ltd
Pan Macmillan, 20 New Wharf Road, London N1 9RR
Basingstoke and Oxford
Associated companies throughout the world
www.panmacmillan.com
www.think-books.com

ISBN 978-1-84525-044-7

Editorial Team: Tania Adams, Claire Ashton Tait, Rica Dearman, Lauren Goddard,
Emma Jones, Simon Maughan, Lou Millward, Richard Rees, Deborah Robertson,
Morag Shuaib, Linda Stanfield, Malcolm Tait, Marion Thompson
Some of the material in this book first appeared in *Wildlife Gardening for Everyone*
Design: Peter Bishop

3 5 7 9 8 6 4 2

A CIP catalogue record for this book is available from
the British Library.

Printed and Bound in Italy by Printer Trento

Visit www.panmacmillan.com to read more about all our books and to buy
them. You will also find features, author interviews and news of any
author events, and you can sign up for e-newsletters so that you're
always first to hear about our new releases.

Main cover image: Blue tit © Tim Gainey / Alamy

THANK YOU

This book would not have been possible without the help and cooperation of the following:

Michael Allen
Helen Bostock
Andrew Halstead
Graham Harrison
Janet Harrison
Stephen Hussey
Keith Martin
Tim McGrath
Pete Mella
David North
Mary Porter
Philip Precey
Mike Russell

Special thanks to Simon Maughan and Morag Shuaib for their time and patience

Special thanks, too, to Laurent Geslin for the use of his excellent photographs.
You can see more of his work at www.laurent-geslin.com

Supported by

SWAROVSKI
OPTIK

Pocket-Sized Long-Distance Viewing

Due to its sophisticated interior focusing system the Pocket 8x20B from Swarovski Optik is the smallest watertight, dustproof, pocket-sized binocular on the market. The world's most intricate optical system, for this category of binocular, guarantees perfect vision. Astonishingly light and compact – the true greatness lies in the detail.

SWAROVSKI
O P T I K

I once had a sparrow alight upon my shoulder for a moment, while I was hoeing in a village garden, and I felt that I was more distinguished by that circumstance that I should have been by any epaulet I could have worn.
Henry David Thoreau

Contents

7

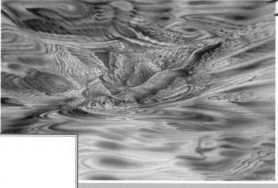

Contents

How to use this book

Although there are some definite dos and don'ts when it comes to gardening for birds, the subject is not always a simple matter of right and wrong. Just as every garden is different, so is every gardener, and each needs to find their own way of sharing their garden with the creatures that also want to live in it.

Available time, space, location, soil type and even simple personal taste are all factors in how each garden will be managed. Everyone will have their own opinions, their own favourite wild visitors, and their own tried and tested approaches to the task.

This book will introduce you to many of these ideas, and help you work out your own favoured approach to bird gardening. Trying out a new type of bird seed; leaving a log pile for insects and seeing what else you invite in; putting up a nest box and wondering which birds might move in – it's all part of the excitement.

As is the sheer anticipation of what you might discover the next time you look out of the kitchen window. No two gardens are the same, and the range of birds that might appear can vary even from one end of the street to another.

Yet there is an approximate pecking order of likelihood. To help you gauge which birds might or might not appear in your garden, we've added a little symbol alongside each of the 60 birds in this book.

And now it's time to get out the binoculars, put out the peanuts, pull up a chair, and settle down to some serious, but highly enjoyable, garden birdwatching.

 Likely visitor

 Possible visitor

 Rare visitor

Flashes of colour, fluttering movement, sweet tunes – whether you have a tiny urban courtyard or a huge field as your back garden, chances are you will see or hear birds there. They'll be looking for food or shelter, or bringing up their young brood at fledging time. It's a great sight, and worth attracting them just to be entertained by their antics.

But there is another good reason to cater for birds – and that is to support them as they face the challenges of coping with threats which include everything from urban development to agricultural intensification to climate change.

Seeing birds in our patch is a reminder that gardens are still a part of nature, a microcosm of what happens out there in the wider countryside. The way we manage our gardens can affect this and the natural world we all share.

It follows that a garden which naturally provides a lot of the food and shelter that birds need will be popular with birds, and what they need is a mixture of insect food and plant food – whether seeds, nuts, berries – or other fruits as well as places to shelter.

Which species are you most likely to see in your garden? Well, the usual suspects tend to be blackbirds, robins, starlings and blue tits. All of these will feed on insects – starlings take leather jackets from lawns and blue tits will help cleanse your roses of aphids. But they will also take seeds (in the case of blue tits) or fruits.

And we can give a little helping hand to the birds that aren't doing too well, such as song thrushes and house sparrows. Song thrush populations have declined by more than half over the past 25 years, and that's enough to give them the maximum listing – Red – as a species of conservation concern. Similarly, house sparrows have declined dramatically and in their case it is thought to be due to low survival rates of chicks, which in turn may be due to low availability of invertebrates. This shows just how important other wildlife in the garden can be to birds.

What about unusual species in your garden? Well, if you live in the South East you may have got used to ring-necked parakeets in your neighbourhood. You will know you have these if you see or hear a flock of chatty green birds passing your way.

Aside from extremely exotic sightings such as sunbirds from south-east Asia and weaverbirds from Africa, as reported to BTO's Garden BirdWatch, you are probably most likely to get unusual birds turning up in winter – for example waxwings can make colourful sightings in some areas. But whether you see them or not may depend on how much food there is in the hedges and fields. Often birds will come into gardens if they cannot find enough food in the countryside.

We also get some species staying on here during the milder winters – for example the blackcap used to be a summer migrant that left again each autumn, but now some of them stay on throughout the winter.

And even though each garden on its own is just one small patch, put them together and you have a huge network of habitats across the country. Get wildlife friendly and you will join The Wildlife Trusts and thousands of people in the UK helping to create green corridors using their back gardens and contributing to a wider landscape movement that helps wildlife adapt to change.

So go on, give it a go. Give wildlife a helping hand. And have fun while you do it.

Bill Oddie

Quickly. Easily. Precisely.

Welcome to the world of digiscoping! It's now possible to capture on camera unique observation experiences. The digital camera base from Swarovski Optik mounts your digital camera directly on to the telescope's eyepiece, simply converting the telescope into a telephoto lens. Photographs with astonishing pinpoint detail can be taken over long distances using your own camera. Recording the moment is simplicity itself thanks to the ability to switch between observation and photography modes in the blink of an eye.

SWAROVSKI
O P T I K

Swarovski U.K. LTD., Perrywood Business Park, Salfords, Surrey RH1 5JQ, Tel.01737-856812, Fax 01737-856885

www.swarovskioptik.com

Wildlife Gardening – an introduction

Matthew Wilson

In 2001 the RHS, The Wildlife Trusts and RSPB organised a joint conference on the theme of Gardens; Heaven or Hell for Wildlife? A range of speakers and delegates were assembled, many of them experts in their respective fields, to consider whether or not the British domestic garden had anything to offer our native wildlife and if so, what? The conclusions drawn were surprising and inspiring; our gardens aren't just good for wildlife, they are vital, and not just the wild and woolly plots that are traditionally associated with wildlife-friendly attitudes.

And what of the legacy of the 2001 conference? Not only has it led to the RHS being even more proactive in encouraging sustainable gardening practices that conserve biodiversity, it has resulted in a hugely beneficial relationship with other organisations, most notably The Wildlife Trusts, with whom the RHS launched the joint Wild About Gardens initiative in 2005.

The Wild About Gardens website provides plenty of practical information for anyone wanting to improve the diversity in their own garden, and a forum for wildlife gardeners to increase their knowledge. Another of the outcomes of the relationship is this book, designed to be practical yet at the same time inspirational and borne out of the knowledge that many gardeners like to attract birds and would like to be more active in gardening with wildlife in mind. At the heart of this book is the advice of bird experts and enthusiasts, keen to pass on the knowledge they have accrued in the best possible circumstances – by doing it themselves.

Gardens form important corridors linking urban, suburban and rural landscapes, allowing wildlife to move from one environment to another. They can also provide tremendous diversity in comparatively small areas, and offer almost limitless habitat opportunities for everything from the smallest bug upwards. In any street in the UK it is possible to find gardens that range from densely planted 'plantsmans' gardens to those bereft of plants altogether, and all stages in between. And it is this diversity which is all important as it provides different opportunities for different species. Gardens can also help to fill in the gaps by providing environments that are otherwise missing locally. So if there are no natural bodies of fresh water in your area a well-designed and planted garden pond can make a huge difference for native pond fauna. Best of all, for plant lovers, is the value that ornamental plants provide for wildlife. Pollen- and nectar-feeding animals get pretty short shrift in winter from our own native flora, but there are plenty of exotic plants that flower during this period and

continued

provide sustenance. Plants that produce berries, nuts and fruit will feed everything from mammals to insects, and of course not only will these exotic plants help our wildlife, they will also fill our days with colour, form and fragrance.

However important the content of our gardens, the way in which we garden is the key to wildlife friendliness. Encouraging natural cycles to develop takes time and patience and can often test the nerves of those who seek perfection in their garden. For many gardeners this can be the area of greatest compromise and, consequently, the hardest to embrace. But consider this: a wildflower meadow is beautiful because we deem it so, not because it is neat, tidy and weed-free. It has an inherent perfection, without being 'perfect'. So we have it within us to redefine what is aesthetically acceptable and to measure any compromise against its value to wildlife. Moreover, the encouragement of natural cycles will help the garden to maintain itself by creating conditions in which natural predators can

thrive. And we can further help the balance in our gardens by avoiding monocultures, using modern disease-resistant cultivars and adopting tried and tested techniques like companion planting.

Now more than ever we need to consider the effect our actions have, and look over the garden fence as well as inside it. Gardens have the capacity to be hugely beneficial to the environment and many gardeners are already enjoying the thrill of seeing their own patch becoming ever more attractive to wildlife. The plants that we use, the habitats we create and the way in which our gardens are maintained have the capacity to preserve and enhance, and surely that must be worth embracing. What could be more satisfying than gardening in a way that not only enhances our own lives, but also that of the biodiversity around us?

Matthew Wilson
Curator and Head of RHS Garden
Harlow Carr

Birds are best

Morag Shuaib, Gardening for Wildlife Officer of The Wildlife Trusts, provides some starting points on encouraging birds into your garden.

For many of us, seeing birds in the garden is the first time we become aware of the wildlife there, and the sheer pleasure of birds' colour, movement and song makes us want to attract them back again. Perhaps we start to see our garden differently: our own backyard becomes a patch of nature that we share with the plants and creatures that arrive there.

The first steps to attracting birds in your garden can include a series of additions to the garden that tackle birds' needs for food, shelter and water – feeders of various kinds, nesting boxes and water baths.

Birds will appreciate all these features, but it is worth remembering that gardens can also provide natural food and shelter, just as much as some wilder habitats. So the way you garden and what you plant in your garden are as important as the artificial devices you may put up.

Your garden is a mini-ecosystem and for some small creatures it is their whole world. Your garden also links up with other gardens around it to form a patchwork of habitats. And because gardens are a whole web of life, many things that you do towards wildlife gardening as a whole will also benefit birds in particular.

What's the best way to keep your garden ecosystem healthy? Some of the most important features are: trees, dead wood, water and variety.

Trees

If you've ever seen blue tits feeding off the mini-beasts on a tree, you'll know that trees can be important for wildlife. And that's just what researchers at Sheffield University discovered. The Biodiversity in Urban Gardens in Sheffield project has shown that mature trees are significant in adding to invertebrate biodiversity in the garden – probably because a mature tree can provide a variety of habitats for all sorts of creatures in its bark, branches and leaves. This in turn will attract the birds.

Dead wood

Dead wood is anything but. It's likely to be full of life and home to many mini-beasts and their grubs. For example, stag beetle larvae can take up to six years to mature and in that time they live in, and feed on, dead wood – so not only is it important to provide a good place for them, the wood also needs to be kept undisturbed.

Why not leave some dead wood standing, or lying about. Create a pile in a shady corner and leave it undisturbed as much as you can.

On a smaller scale, you can also make insect nests – or an insect-hotel – by gathering some twigs and tying them together. Leave them horizontally and they may attract insects looking to lay their eggs, or looking for an overwintering site. Leaving your perennial plants standing throughout the winter will also provide shelter for lots of insects – and more food for birds the following spring.

Water

Water is important to birds not just to drink, but also to help them keep their feathers in good shape. You can provide water either with a birdbath, or as a shallowly sloping edge of a pond. A pond has the added bonus of adding insect life to the garden and so a greater variety of food sources for birds.

Variety

Variety in your garden planting won't just make your garden more interesting for you – it will also provide different features for 'your' birds. Birds need structural variety from trees, shrubs, climbers and perennials, which will give them shelter and breeding places as well as perching spots and safety from predators. They also need a variety of food sources – from seeds to berries, and nuts to insects. The longer you can extend your flowering and fruiting season the better.

So if you have dead wood and insect hotels in your garden you are well on the way to attracting some of the insects and mini-beasts that birds like to eat. If you have a garden full of nectar plants, this will help attract insects that birds will feed off. If you have plants that fruit (whether as berries or larger fruits such as apples) – or that set seed (as with grasses and perennials and annuals) then seed-eaters will be happy.

Finally, if you chance to see a sparrowhawk or other flying predator diving into your garden and catching a smaller bird it may be upsetting, but this is one other strand in the web of life, and no less worthy...

Plants for seeds
- Angelica (*Angelica sylvestris*)
- Clematis (*Clematis* spp.)
- Globe thistle (*Echinops ritro*)
- Goldenrod (*Solidago canadensis*)
- Greater knapweed (*Centaurea scabiosa*)
- Honesty (*Lunaria annua*)
- Meadowsweet (*Filipendula ulmaria*)
- Sunflower (*Helianthus annuus*)
- Teasel (*Dipsacus sylvestris*)
- Yarrow (*Achillea* spp.)
- Grasses

Plants for fruits
- Barberry (*Berberis* spp.)
- Bramble (*Rubus spp.*)
- Elderberry (*Sambucus nigra*)
- Guelder rose (*Viburnum opulus*)
- Holly (*Ilex aquifolium*)
- Japanese quince (*Chaenomeles spp.*)
- Oregon grape (*Mahonia spp.*)
- Pyracantha (*Pyracantha spp.*)
- Spindle (*Euonymus europaeus*)
- Yew (*Taxus baccata*)
- Fruit trees – for example, apple and pear – fallen fruit will provide a feast in the autumn

Birds are best

continued

Early season nectar plants
- Aubretia (*Aubretia*)
- Flowering currant (*Ribes sanguineum*)
- Grape hyacinth (*Muscari botryoides*)
- Lungwort (*Pulmonaria spp.*)
- Primrose (*Primula vulgaris*)
- Sweet violet (*Viola odorata*)
- Wallflower (*Erysimum cheiri*)
- Winter aconite (*Eranthis hyemalis*)
- Wood anemone (*Anemone nemorosa*)
- Yellow alyssum (*Alyssum saxatile*)

Mid-season nectar plants
- Buddleia (*Buddleja davidii*)
- Heather (*Calluna vulgaris*)
- Lady's bedstraw (*Galium verum*)
- Lavender (*Lavendula spp.*)
- Mallow (*Lavatera spp.*)
- Purple toadflax (*Linaria purpurea*)
- Rock cress (*Arabis caucasica*)
- Sea holly (*Eryngium maritimum*)
- Verbena (*Verbena bonariensis*)

Late-season nectar plants
- Coneflower (*Echinacea purpurea*)
- French marigold (*Tagetes spp.*)
- Golden rod (*Solidago canadensis*)
- Honeysuckle (*Lonicera spp.*)
- Ice plant (*Sedum spectabile*)
- Ivy (*Hedera helix*)
- Meadow saffron (*Colchicum autumnale*)
- Michaelmas daisies (*Aster novi-belgii*)
- Perennial sunflower (*Helianthus spp.*)
- Red valerian (*Centranthus rubra*)

Evening nectar plants
- Evening primrose (*Oenothera biennis*)
- Greater stitchwort (*Stellaria holostea*)
- Night-scented stock (*Matthiola longipetala*)
- Tobacco plant (*Nicotiana spp.*)
- White campion (*Silene latifolia*)

Herbs
Herbs are great for attracting a variety of insects in the garden, such as day-flying moths and hoverflies.
- Angelica (*Angelica spp.*)
- Borage (*Borago officinalis*)
- Catmint (*Nepeta spp.*)
- Chives (*Allium schoenoprasum*)
- Fennel (*Foeniculum vulgare*)
- Hyssop (*Hyssopus officinalis*)
- Mint (*Mentha spp.*)
- Rosemary (*Rosmarinus officinalis*)
- Thyme (*Thymus spp.*)
- Wild marjoram (*Origanum vulgare*)

Birds are best

continued

Butterfly and moth larval food plants

We love to see butterflies, but we would not have them without their caterpillars So why not try planting some of the following (although be aware that they do not all work all of the time, as a BUGS project for Sheffield University recently discovered with nettles):

- Alder buckthorn – brimstone
- Birdsfoot trefoil – dingy skipper and burnet moths
- Common nettle – red admiral, comma, peacock and small tortoiseshell
- Cuckoo flower – green-veined white
- Garlic mustard – orange-tip
- Grasses – speckled wood
- Holly and ivy – holly blue
- Rosebay willowherb – elephant hawkmoth
- Thistles – painted lady

Are you Wild About Gardens ?

If you enjoy reading this book and would like to know more about wildlife gardening, why not log on to the Wild About Gardens website at www.wildaboutgardens.org. Wild About Gardens is a joint project of The Wildlife Trusts and the Royal Horticultural Society, and aims to inspire people to take action for wildlife through gardening. On the website, you'll find sources of information on all sorts of things such as having more wildlife-friendly flowers in your borders, building garden ponds, and what berries and seeds are good for wildlife. You could also have the chance to join in one of our surveys, you can read about other gardener's experiences, and you can join in by sharing your own stories, experiences and observations.

About The Wildlife Trusts

If you want to get involved with wildlife-friendly gardening, The Wildlife Trusts are the ideal place to go for advice. Whether you're after tips on how to build a pond or where to put a nest box, your local Trust will be able to help you attract wildlife to your garden. The Wildlife Trusts also run the Wild About Gardens project – in partnership with the RHS – developing the links between the worlds of gardening and nature conservation for the benefit of people and wildlife.

There are 47 local Wildlife Trusts across the whole of the UK, the Isle of Man and Alderney, which are working together to fulfill our collective vision of 'an environment rich in wildlife for everyone'. With over 720,000 members we are the largest UK voluntary organisation dedicated to conserving the full range of the UK's habitats and species. Collectively, we manage more than 2,200 nature reserves spanning over 80,000 hectares. But our work resonates far beyond these sites. Each year we work with more than 5,000 land owners, and give over 10,000 days of advice on wildlife-friendly practices. We also campaign for the protection of wildlife and invest in its future by helping people of all ages gain a greater appreciation and understanding of wildlife.

Our work with wildlife is not restricted to the countryside either. Some of The Wildlife Trusts' most important projects are in the heart of our towns and cities, whether it be organising the London Stag Beetle Hunt or helping people to nurture community space in urban Sheffield.

Engaging people is core to our work and we pride ourselves on working with a diverse range of communities. We hold over 6,000 walks, talks and events each year and are fortunate to be supported by more than 33,000 volunteers. As well as inspiring people about the wildlife around them, we also strive to spread the message about the need for our society to reduce its impact on the world about us.

Involving the next generation is all-important, and each year we work with more than 12% of UK schools, helping children to learn more about the natural world. We have over 100,000 junior members in 'Wildlife Watch', all actively learning about our plants and animals while having fun and making new friends.

Our work doesn't stop on land. The Wildlife Trusts also strive to protect that last great UK wilderness and home to more than half our species life: the sea. Vitally, we are campaigning for designated, highly protected marine reserves, like the reserves we have on land.

Our terrestrial nature reserves have long been key to the protection of UK land-based wildlife. Yet how many people know that in the UK our private gardens, covering about 270,000 hectares, are larger collectively than all the National Nature Reserves? We know about the threats to our wildlife, but there are also great opportunities as well. The more wildlife-friendly gardens we have, the stronger the UK's wildlife network becomes: that system of green refuges and corridors which our native species rely on for food, movement and habitats.

Gardeners play a vital role in The Wildlife Trusts' vision for landscape-scale conservation – helping us to stitch the countryside back together, rebuild biodiversity and let wildlife adapt to threats such as habitat loss and climate change. To read about the Living Landscape vision log on to www.wildlifetrusts.org and click on 'publications' then 'free publications'.

Whether in gardens or beyond, the spirit of The Wildlife Trusts is embodied in local action. So what could be simpler, or closer to home, than creating a haven for wild birds in your garden?

Nuts about wildlife?

THE wildlife TRUSTS

join us today
for
- *Natural World* magazine three times a year
- Local Wildlife Trust newsletter
- Invitations to events, local to you
- The knowledge you are helping to protect wildlife!

Individual: £24 • Joint: £30
Family: £36*(incl. children's club Wildlife
Watch, up to 4 children at the same address)

About the RHS

The UK's leading gardening charity

The RHS believes that horticulture and gardening enrich people's lives. We are committed to bringing the personal and social benefits of growing plants to a diverse audience of all ages, to enhance understanding and appreciation of cultivated plants, and to provide contact with the natural world.

We believe that good practice in horticulture and gardening in both private and public spaces is a vital component of healthy sustainable communities and the creation of long-term environmental improvements.

The RHS has four gardens around the UK: Wisley in Surrey, Hyde Hall in Essex, Rosemoor in Devon and Harlow Carr in North Yorkshire, as well as over 100 partner gardens. It runs some of the best flower shows in the world and is a world leader in horticultural science and libraries. The RHS runs education and outreach programmes across the UK to help people realise their potential. Its charitable activities include:

- Helping over 600,000 children a year through hands-on opportunities for children to grow plants and learn about the natural environment
- Encouraging sustainable horticulture and environmental care
- Promoting biodiversity and creating wildlife habitats at its gardens
- Providing expert gardening advice
- Running the RHS Award of Garden Merit scheme to help gardeners choose plants that perform well
- Undertaking scientific research into issues affecting gardeners
- Educating and training gardeners of all levels and experience
- Maintaining the Lindley Library, the finest horticultural library in the world

The RHS relies heavily on donations and sponsorship to help enhance people's lives through gardening. For every pound received from its members' subscriptions, the charity needs to raise more than twice as much again to fund its charitable work. While the charity receives some income from garden operations, flower shows, shops and plant centres, support from donors and sponsors is vital to its mission to bring the benefits of gardening to all.

Royal Horticultural Society

Join the RHS today

JOIN TODAY & SAVE £5

PLUS FREE £5 SEED VOUCHER WHEN YOU JOIN BY DIRECT DEBIT

ENJOY THE BENEFITS OF RHS MEMBERSHIP:

- **Free entry** with a guest to RHS Gardens: Wisley, Rosemoor, Harlow Carr and Hyde Hall

- **Free access** to over 140 RHS Recommended Gardens throughout their opening season, or at selected periods

- **Free monthly magazine**, *The Garden*, full of practical advice, ideas and inspiration, delivered to your door (*RRP. £4.25*)

- **Privileged entry and reduced rate tickets** to the world famous RHS Flower Shows: Chelsea, Hampton Court Palace and the RHS Flower Show at Tatton Park

- **Free Gardening Advice Service** available by post, telephone, fax or email all year around

Take advantage of the £5* saving and pay only £39 when you join today.

Call **0845 130 4646** and quote 1998
(Lines are open 9am – 5pm, Monday to Friday)

or join on line at **www.rhs.org.uk/joinpromotions**

TERMS AND CONDITIONS Offer is only available on Individual (£44) and Family membership (£68). Offer is available until 31.10.07. This offer cannot be used in conjunction with any other offer.
* Individual membership is £44 and includes a one-off £5 enrolment fee.   Registered Charity No. 222879

The RHS, the UK's leading gardening charity

What a picture!

Enter the RHS Photographic Competition, and your images could soon be appearing in books like this.

You don't have to be a professional to get your photographs published. Many of the images in this book were taken by amateur naturalists and enthusiastic gardeners who simply love to record the wild world around them. And some of them are real winners.

The aim of the RHS Photographic Competition is to convey the diversity of plants and the enjoyment of gardening through the medium of photography. The competition is open to amateur and professional photographers and the growing popularity of the competition has seen entries rise to over 4,000 images. The competition 'Photograph of the Year' categories enable photographers to explore opportunities with Plant Portraits, Close-Up, Trees or Shrubs, Wildlife in the Garden and People in the Garden.

For more details please contact RHS Garden Rosemoor: Tel: 01805 624067
Web: www.rhs.org.uk/news/photocomp.asp
Post: RHS Garden Rosemoor, Great Torrington, Devon EX38 8PH

Photo credits

The photographs in this book are copyright RHS, The Wildlife Trusts, or the people and organisations listed. We would like to thank them for making this book possible:

Neil Aldridge (p204), www.ardea.com, Richard Burkmarr (pp 33, 43, 51, 70, 71, 73, 77, 79, 86, 88, 89, 93, 94, 95, 97, 109, 167, 179, 181, 210, 212, 224), Isabel Carrahar (49, 99, 191, 193), Alison Clarke (16), www.enviromat.co.uk (91), Gatehouse Studio (199), Gloucestershire Wildlife Trust (115), Hampshire and Isle of Wight Wildlife Trust (125), Jerry Kavanagh (64), Ray Kennedy (rspb-images.com), Paul Marten (172, 173), Keith Martin (205), David Plummer (157), Maddalena Saccon (170), Morag Shuaib, www.shutterstock.com, Sussex Wildlife Trust, Phil Sutton (69, 125, 199, 208, 209), Sue Tatman (208), Olive Tayler (117), Scott Tilley (113, 115), Ulster Wildlife Trust (147, 208, 213), Neil Wyatt (208, 223), Wildstock (27), Tim Webb (23).
Some of the photographs in this book were provided by entrants to RHS photo competitions, and are copyright the owners. Our thanks to Dr S Venitt (87) and Mike Calvert (133).

Special thanks to Laurent Geslin, www.laurent-geslin.com

Blackbird
Turdus merula

Of all the birds that might call your garden home, the male blackbird is about the easiest to identify. He is literally the glossy black bird, with a bright golden beak. Although the female shares the male's chunky physique and longish tail, she looks more like the chocolate bird. Their bossy behaviour around the garden also gets them noticed. When either of them spots another blackbird trespassing on its territory, it swoops in, issuing a scolding *chik-chik-chik* as it lands, tail cocked and wings drooping, and goes on the warpath to see off the intruder.

While a pair of blackbirds has a brood of chicks to feed, you often see the parents running and hopping over the lawn, scouring the short grass for earthworms and insects. Occasionally, one stops, head tilted to stare at the ground, before plunging its beak into the grass and tugging out a worm, then flying back to the

Did you know?

■ After gathering a bunch of dry grass, some canny female blackbirds fly to a sticky puddle and smear the hay in mud before pasting it into their nests. To make a blackbird's task of finding nesting material easier, you could leave piles of moss and straw around and create a muddy patch in a flowerbed.

■ If you find half a greenish-blue eggshell with chestnut freckles lying on the lawn, you know there's a blackbird nesting in the vicinity, but not necessarily close by, as the parents drop the shells well away from the nest in order not to give its location away.

■ White blackbirds turn up quite frequently, especially in towns. A few are all-white, but retain their yellow beaks and eye-rings so are not true albinos; most just have patches of bleached feathers on the head, wings or tail. Such whiteness seems to be caused by a combination of breeding, diet and disease.

Facts

Frequency
A widespread, familiar garden bird. About five million pairs breed here, but numbers swell to 10 to 15 million after the breeding season and when British residents are joined by flocks from Scandinavia and northern Europe for the winter.

Identification
The male is an all-black bird, embellished only by a golden beak and eye rings; the female is an earthy-brown bird – with her paler throat, lightly-speckled chestnut breast and brown beak, she may be mistaken for the speckled song thrush; juveniles are a redder and more spotted version of the female.

Song
Described as 'the Beethoven of bird music', as the male sings exquisite musical phrases in mellow flute-like tones, with apparent ease.

Nesting
The female selects a safe, concealed location, often in dense creeper against a wall or fence, or in leafy trees or bushes, where she fashions a deep cup of entwined roots and moss, bound with a lining of mud and softened with a layer of dry grass.

Length
24-25cm (9in)

Conservation
Not in immediate trouble; garden feeding and nesting are making up for a loss of natural habitat due to changes in farming practices.

nest with a beakful of squirming food for the family. Sometimes you will also hear a blackbird rummaging vigorously through the leaf litter, looking for small invertebrates to eat.

Even when you can't see any blackbirds, you're bound to hear them; they loudly cackle *chook-chook-chook* as they fly off after being disturbed, and have the last word as night falls and they retire to roost, accompanied by noisy *tic-tic-tic* calls. In spring and summer, the male blackbird wakes before sunrise to lead off the dawn chorus and is often still singing his melodic song from the top of the tallest tree or post, long after the sun has gone down.

A woodland bird, the blackbird appreciates the cover and food provided by garden shrubs and hedges. No sooner have berries begun to ripen and turn red on trees and bushes, and in hedgerows, than blackbirds are snaffling them: starting with cherries, then rowan, before honeysuckle and hawthorn. Only when they start pilfering raspberries and strawberries do some gardeners get a bit fed up with them. Later in the year, blackbirds tuck into windfall apples, and scraps and dried fruit on the bird table, to supplement their dwindling hedgerow harvest.

**'The blackbird's mellow fluting notes
Call my darling up with round and
roguish challenge:
Quaintest, richest carol of all the
singing throats!'**

From 'Love in the Valley' by George Meredith

weasdale
nurseries

Growers of bare-root trees & shrubs at 850ft elevation since 1950. Mail-order our speciality.

The Weasdale Native Wildlife Hedge-mix is carefully selected to mimic the plant species diversity of a typical centuries-old country hedge and will come to host a wide range of insects, animals and birds to bring added colour and life to your garden.

Measure the length of your intended hedge and let us quote you for sufficient plants to create your own piece of the country in your back garden.

Alternatively, send £2.00 (2007-08) for our detailed, highly readable, illustrated and unputdownable catalogue which lists the hundreds of native and ornamental trees & shrubs that we grow at our Cumbrian nurseries in the Howgill Fells.

All this and much more can be found on our website at **www.weasdale.com** - pay us a visit there today.

KIRKBY STEPHEN (WGB) . CUMBRIA . CA17 4LX

tel 015396 23246   fax 015396 23277

sales@weasdale.com   www.weasdale.com

Pete Mella, Sheffield Wildlife Trust, advises:

■ One of the most common and recognisable birds in Britain, the blackbird is a familiar sight in all but the most upland of areas. Although it's primarily a woodland species, it's a very common visitor to parks, gardens and hedgerows, and is found throughout the British Isles.

■ The blackbird feeds mainly on insects and berries and will eat earthworms throughout the year, as long as the ground is soft enough. It's a regular sight to find a blackbird in your garden immediately after gardening, looking for dug-up worms. During autumn and winter it will also eat ripe fruit, such as fallen apples, and the presence of these can attract them to your garden. They will readily feed from a ground feeder, taking foods such as raisins and kitchen scraps.

■ Blackbirds are generally solitary birds, and will noisily defend their territories. They make a cup nest out of grass, twigs and mud, which is usually made in a hedge or bush, but can be on shelves in outbuildings. The majority of nesting attempts fail, as their open nests are vulnerable to weather conditions and predation.

■ Although always-common birds, the last 25 years have seen blackbird numbers suffer a slow decline of around 20%, and at one point they were moved on to the Amber Conservation List. The areas hardest hit were on farmland: this may have stemmed in part to government agricultural policies that led to the removal of hedgerows. This decline is in reverse now, and the blackbird is back on the Green List of least-concern species.

■ A little-known fact is that our blackbird numbers swell during the winter months, as our resident birds are joined by birds from Scandinavia, and even as far east as Poland and Russia.

Blackcap
Sylvia atricapilla

Blackcaps discovered the attractions of garden feeding relatively recently. They were once a definite tick in the summer visitor box, arriving back from their wintering grounds in Spain, Portugal and North Africa in March or April before setting off south again in the autumn. But lately, more of them have been over-wintering in this country too, mainly in the south, west and Midlands. A few are hardy residents that stay here all year, although many breeding in Britain still undertake the long pilgrimage to warmer climes. But as they leave, some of the blackcaps that spent the summer nesting or being reared in Scandinavia and northwestern Europe start arriving on these shores for the winter.

Wintering in the UK has its advantages and risks for blackcaps. On the plus side, shorter and less hazardous return journeys to their breeding grounds in the UK each spring gives them a headstart on further-flung migrants, because they get back earlier and have the pick of the best territories. Research suggests that birds over-wintering in the British Isles produce larger clutches and more young than those migrating further south. But as an insect-eater, the blackcap gambles on the weather staying mild and being able to find enough food. The fact that the blackcap is more vegetarian than other warblers, and can exploit bird tables, has helped a growing number survive the British winter.

Most blackcaps disappear into deciduous woodland and hedgerows over the spring and summer months. There they build their nests in brambly undergrowth and nettle patches, and track down insects and spiders among the foliage. For bird lovers, the biggest bonus of the blackcap's changes in travel plans and winter destinations is that they have become more visible in the garden from autumn onwards. They forage for insects around compost heaps and eat ivy, cotoneaster, holly, elder and mistletoe berries. Many now also visit bird tables, taking grated cheese and suet, chopped peanuts, scraps, breadcrumbs and pinhead oatmeal.

'And in the little thickets where a sleeper
For ever might lie lost, the nettle-creeper
And garden warbler sang unceasingly'

From 'Haymaking' by Edward Thomas
('nettle-creeper' is an old country name for the blackcap)

Facts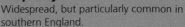

Frequency
Widespread, but particularly common in southern England.

Identification
Quite chunky as warblers go, but less anonymous than most because the male and female have differently coloured skullcaps: his is black, hers is brown; he has pale ash-grey underparts, a browny-grey back and grey wings and tail; she has a browner tinge all over; the juvenile looks like the female.

Song
A rich and varied warble. Also a sharp *tic* call.

Nesting
Blackcaps build their nests quite low to the ground, generally in brambles and undergrowth.

Length
13.5-15cm (5-6in)

Conservation
Seems to have taken matters under its own wings; so far, swapping winters in Iberia and Morocco for life in British gardens is showing a better survival rate, which has led to a larger breeding population.

Did you know?

■ There's a lot of speculation about what induced more blackcaps to tough out the winter here. The first blackcaps to do so might have been en route to North Africa from Scandinavia and, instead of passing through Britain on their way south, ended up staying until the following spring, when they headed north again. Or it is possible that a few blackcaps nesting in eastern Europe in the 1950s suffered a navigational glitch and instead of flying south, flew westwards and ended up in Britain. Lost but lucky, they managed to survive a mild winter and they, too, retraced their wingbeats in the spring. And, so the theory goes, their descendents have been following the same route ever since.

■ One unexpected consequence of blackcaps staying in Britain for the winter may be more sprigs of mistletoe to hang up at Christmas. The blackcap is partial to mistletoe berries: after eating the fleshy bits, it strops its beak across a nearby branch to knock off the seeds stuck to it with the sticky berry juice.

Andrew Upton, Ulster Wildlife Trust, advises:

■ Blackcaps are largely summer visitors to Britain and Ireland, and winter in southern Iberia and northwest Africa. They breed in a wide range of woodland and shrubby habitats. The British population is currently increasing and expanding its distribution further north and west.

■ It appears that birds wintering in Britain and Ireland are distinct to the breeding population and originate from Germany, Belgium and the Netherlands.

■ Increasing numbers of blackcaps have been seen in gardens in winter over the past 30 years, assisted by our milder climate. They are attracted to gardens to feed on shrubs loaded with berries and seed provided at bird tables.

Blackcaps are usually seen on their own or in pairs.

■ Encourage blackcaps to feed in your garden by planting shrubs that provide a wealth of berries throughout the winter, such as ivy, hawthorn, *Pyracantha* and cotoneaster. And remember to provide a good mix of bird seed on bird tables and drinking water to help them through our winters.

Black-headed Gull
Larus ridibundus

When is a seagull not a seagull? When it's a black-headed gull. This small gull is much more of an inland bird than a sea-faring one, and is rarely seen far off shore. If you live near a lake, waterway or gravel pit, you're likely to see them flying overhead. It's unusual for them to come into gardens, though, unless they're fairly large, and there are some good pickings to be had. Black-headed gulls are carrion eaters and will pick over rubbish tips to see if some tasty morsels are likely to come to light. They're agile gulls, too, able to pluck flying ants out of the air while on the wing.

One of the reasons that these birds don't tend to go for gardens is that they like to hang around in groups. Gangs, though, is probably a better word. A flock of black-heads is a noisy collection, calling and quarrelling and generally making something of a nuisance of themselves. If you do see some close up,

however, you'll discover that they're not particularly well named. For much of the year these birds have white heads with small dark markings behind the eyes, and come the breeding season the colour of their heads becomes more chocolate than black, although this colour does not cover the bird's nape. There's another British gull, occasionally seen in flocks, called the Mediterranean gull which has a much blacker head.

Black-headed gulls have adapted well to human surroundings, and have become common sights in towns and cities up and

'In ritual circles, resolute and high,
He smoothly, purely passed.
I almost heard
The music which that sun perfected bird
Left in the soundless sky.'

From 'The Gull' by Frances Cornford

Facts

Frequency

Britain's commonest inland gull, some 200,000 pairs breed in the UK annually, and are joined by well over a million more from the continent during the winter months. It can be found across Britain and Ireland.

Identification

A smallish gull with a chocolate brown head during the breeding season, dark red bill and red legs. The white leading-edge to its wing is diagnostic. Young birds are gingery brown on the upperparts, these markings turning into bars by the first winter. The following year the young have a semblance of a dark head, but the full colour does not develop until the second summer.

Song

A scolding *keehar*.

Nesting

Black-headed gulls are ground nesting birds inland, finding suitable sites on islands in rivers, lakes or gravel pits. They sometimes also lay their eggs in large reedbeds.

Length

35-38cm (14-15in)

> **Outside of the breeding season, the best way to identify this gull is by the white leading-edge to its wing.**

down the country. Parks, playing fields and, of course, rubbish dumps are all likely spots to find them, then as evening falls they make their way back to their communal roosts on lakes and reservoirs. It's in the winter that they're most likely to come into gardens, once their numbers are swelled by continental arrivals, and competition rises. If you forget to put the lid on your wheelie bin, or you have an exposed compost heap, there's a chance that some of them might make an appearance, yelling at anything that gets in their way.

When you see them on water, they adopt a rather comically alert position, their tails raised at quite an acute angle, and their necks stretched long.

Did you know?

■ Black-headed gulls are far-flung birds, breeding not just in Europe but on the eastern coast of Canada and in Asia, too.

■ The scientific name of this bird means 'laughing gull', which makes sense when you hear the ribald *keehar* call rattling around a colony.

■ The black-headed gull takes two years to reach maturity.

Blue Tit
Parus caeruleus

Did you know?

■ Blue tits prefer to nest in tree holes, but suitable sites are often in short supply and oversubscribed. You can help them find a suitable nesting place and encourage them to nest in your garden by putting up nest boxes. To minimise competition for nesting space, pick out ones with a 25mm entrance hole, which the blue tit can use, but great tits, house sparrows and starlings are too large to get through.

■ Apparently, blue tits have agile brains as well as bodies. They soon work out how to solve puzzles, such as one which involved removing matches to shunt a peanut on towards the end of a maze where the blue tit could claim its reward by eating the nut.

■ Not so long ago, cute little blue tits were notorious doorstep vandals. Back in 1929, a blue tit in Southampton was the first to be spotted pecking at a shiny foil milk-bottle top, apparently to get at the creamy top-of-the-milk underneath. The habit soon spread through the blue tit population, probably as one bird copied another in winter flocks. Within 20 years, blue tits were sabotaging morning pintas on doorsteps all over the country. But as quickly as they took up the idea, they stopped, probably as house-to-house deliveries declined and skimmed or semi-skimmed milk became more popular, or maybe simply because they were thwarted by a change from bottles to cartons and plastic containers.

As the nimblest member of the garden acrobatic display team, the blue tit is an endearing and colourful little favourite. Its animated, sure-footed tumbling on bird feeders and along the slenderest twigs is endlessly distracting. What at first seems to be inquisitive but haphazard fluttering and somersaulting over branches and leaves, or wanton destruction of blossom on a tree, turns out to be an urgent search for tiny insects and spiders concealed in bark or among petals.

Finding enough caterpillars, moths, aphids and other tiny creepy-crawlies to eat occupies most of a blue tit's time. Over the course of a year, blue tits devour huge numbers of garden pests: among their favourite foods are codling moths, green oak tortrix moth caterpillars, sawfly maggots and weevil grubs. In winter, when insects are harder to find, blue tits have come to rely on seeds and nuts put out on bird tables or in bird feeders.

In April and May each year, a pair of blue tits devotes its energy to raising a single large brood – 10 eggs are not unusual. Nesting coincides with a population explosion of tree caterpillars, so the parents can find plenty of food fast enough to satisfy so many hungry beaks. Anger or excitement, especially around the nest, brings out a blue tit's feisty, fearless nature. Up go the blue feathers on the crown of its head into a shallow crest, while its tail and wings start flicking as it lets rip with a diatribe of churring and hissing.

Sadly, the blue tits' parental dedication doesn't always have a happy ending. Blue tit fledglings face a steep learning curve to master flying and survival skills after leaving the nest in late May. Even though their parents continue to feed them for a week or so, the naive youngsters are vulnerable to predation and bad weather. Pet cats are some of the worst offenders when it comes to killing young blue tits. The harsh reality is that, on average, nine out of 10 chicks are likely to die before the following spring.

Facts

Frequency
The commonest tit in the British Isles; numbers show a steady increase, thanks largely to more garden feeding and the provision of nest boxes.

Identification
The blue skullcap and yellow breast are fairly easy to follow as the bird flits between twigs on short, rounded wings; juveniles are fluffier, with greyer and paler yellow plumage.

Song
A jolly trilling *tsee-tseee-tsu-chu-chu*.

Nesting
The male picks out a nest site in a hole in a tree or wall, or a nest-box, where the female builds a cosy nest-cup from moss and dry grass which she stuffs with feathers and hair.

Length
11-12cm (4-5in)

'Where is he, that giddy Sprite, Bluecap, with his colours bright, Who is blest as bird could be, Feeding in the apple tree'

From 'The Kitten and Falling Leaves' by William Wordsworth
('bluecap' was a popular name for the blue tit in the 19th century)

WOLF Garten

It's the multi-change® silver jubilee

25 Years · multi-change® · WOLF Garten · **click**

"changing tools in a click"

This year heralds the 25th anniversary of the multi-change® tool system from WOLF-Garten, whereby any tool head from the range snaps onto any of the handles. Customers who have experienced the many advantages of the multi-change® »click« solution don't want to change systems ever again. Our premium quality, ergonomic tools are manufactured in Germany and cater for every type of gardener and every type of garden.

Visit **www.WOLF-Garten.com** or call **01495 306600** for a full list of stockists.

10
Jahre Garantie
Year Guarantee
Anni di Garanzia
Jaar Garantie
Ans de Garantie

Janet Harrison, Warwickshire Wildlife Trust, advises:

■ The blue tit is a familiar, endearing garden bird that readily comes to garden feeders. Partial to peanuts, it soon learns how to extract the kernels from shells hanging on a string, or empty a container of shelled nuts. However, use unsalted nuts, as salt is disastrous for most birds. Blue tits also favour seeds, particularly sunflower seeds, bird puddings and fat balls. You can now buy special containers to put the fat balls and other food in.

■ Remember to keep your feeders extremely clean to avoid diseases passing between birds.

■ Blue tits also take natural food, both seeds and insects, from the garden, so trees, bushes and herbaceous plants also attract them. Extremely agile, these acrobatic little birds often feed upside down, even on the most slender birch twigs, in search of the nest meal. Insects, especially caterpillars, are essential food for nestlings, providing their only source of fluid, so try to avoid using insecticides in your garden.

■ With milder winters and more people providing food, numbers of blue tits are increasing. But natural nest holes can be in short supply, so nest boxes are regularly used. If necessary, protect the hole with a metal plate to stop birds such as woodpeckers from enlarging it. Boxes should face north-eastwards to prevent overheating and should not have a perch or nearby branch that could be used by predators, such as squirrels and cats. The nest boxes should also be cleaned once the young have left and the breeding season is over.

Brambling
Fringilla montifringilla

If you live in a rural area you have a reasonable chance of spotting a brambling or several in your garden. But it all depends on the year. Bramblings are migratory finches that arrive each September from Scandinavia, but their numbers are extraordinarily variable. Some years up to a couple of million will make their way to this country; in others there may be no more than a few tens of thousands. This extreme variability is all down to availability of food supplies. Flocks that appear in Britain one year have been recorded in Italy the next.

Perhaps for this reason, bramblings are among the most overlooked of British birds.

With no set patterns to go on, it's difficult to know when to look out for them. But when they do arrive in numbers, they can make quite an impression. The sight of possibly thousands of the birds toing and froing across fields in search of precious grain, often in the company of their close relatives the chaffinches, is most endearing.

Not that you're likely to see them for long. Timid birds, they'll flick away from you at the drop of a hat, leaving you with a glimpse of their retreating white rump, a key feature in picking them out from their southern counterparts.

Fields alongside beech woodland are the best places to see these birds, and if you live alongside

Facts

Frequency
This bird winters in the UK throughout the country apart from the far north and west of Scotland. The birds make Britain their home in the winter so the best time to look out for them is from mid-September to April.

Identification
The brambling is a finch whose plumage changes quite considerably from season to season. In the winter, while it's in the UK, it has a streaked head, orange breast and white rump, and its upper parts are mainly black but mingled with orange. The female's colours are less intense. In the spring, the male's head becomes much darker and their full-blown breeding plumage, with its rich orangey-red blush on the wing and breast, are revealed. Unfortunately for Brits, the bird has normally returned home by then.

Song
A soft *chuck chuck* sound is followed by a rasping *zweek*.

Nesting
The nest will not be seen in Britain as the bird leaves the country for the summer, but it is built by the female, usually in a conifer. It's a deep cup shape made from moss, bark and hair, and lined with wool and feathers. It is decorated with bark and lichen.

Length
14-15cm (5.5-6in)

such a habitat then you might want to prepare your bird table with beech seeds to try to tempt a few in. Shelled, flattened peanuts have also proved themselves to be good enticements.

Did you know?

■ Very occasionally, bramblings will stay on into the summer months and breed in Britain, but this happens only about half a dozen times in a year at the most.

■ Bramblings are birds that have suffered from the caged bird industry of the past. Bird-catchers in Yarmouth were particularly destructive to local populations.

Bramblings are highly migratory finches, but their migratory paths are most irregular, depending upon the availability of beechmast and other favoured seeds.

Bullfinch
Pyrrhula pyrrhula

For a strikingly handsome bird with such a brilliant pink breast, the male bullfinch keeps a remarkably low profile. Clear views are few and far between, partly because bullfinches rarely venture far from the leafy cover of thickets, copses and hedgerows, but also because there are not many of them around these days. More often than not, it's a flash of the distinctive square patch of white rump above the black tail flying along a hedgerow or into cover that triggers a bullfinch alert.

Watch closely and you might catch a glimpse of the male's jet-black skullcap and beak, grey back and his bright-pink cheeks, throat and breast. Depending on the light, the colour can vary from salmon pink and brick, or strawberry red, to the deep pink of horse-chestnut blossom; in bright sunlight it can look almost blood red. When you see a male bullfinch, there's a high probability you'll see a female tagging along with him, because they stay together as a pair

Did you know?

■ The bullfinch has long been officially recognised as a pest. Way back in the 15th century, it carried a bounty on its head. In the 1950s and 1960s, bullfinch numbers increased so greatly that the fruit growers in southern England were granted licences by the Ministry of Agriculture, Fisheries and Food to cull them. Using live-decoy baited traps thousands of the birds were killed each year. But it wasn't until the 1970s that numbers started to fall alarmingly, although the authorities were slow to acknowledge the bullfinch's predicament. Despite appearing on the Red List of British birds at serious risk of vanishing all together, and being protected by the Wildlife and Countryside Act of 1981, it remained legal to trap and destroy bullfinches until 1996. Amendments to the licensing law now permit culling only if there is evidence of serious damage, and after non-fatal methods of preventing damage by spreading fine netting or threads over trees have proved ineffectual.

■ Bullfinches display an unaccountable preference for the buds of certain varieties of fruit tree, attacking the buds of James Grieve apples, Morello cherries, Conference or Williams pears and greengages, but turning their beaks up at Bramley apple and Comice-pear buds.

throughout the year, or travel about in small family flocks over the winter.

Bullfinches may be colourful birds and loyal partners, but they have long had a demon reputation for denuding fruit trees and bushes of their buds in spring, and many commercial fruit growers – and some gardeners – despise them for it. For most of the year, bullfinches scour hedgerows and woodland for seeds and berries on wildflowers and trees. They're especially fond of ash samaras, the bunches of winged seeds dangling on ash trees. Damage to fruit trees in orchards and gardens in years when ash trees seed heavily is less severe than in alternate ones when they take a rest from seeding.

From December onwards, as wild seeds and fruits become scarcer, hungry bullfinches start eating the buds containing the future leaves and flowers of cultivated fruit trees and bushes in gardens and orchards (their buds start swelling before the ones on wild trees). Individually, each bud isn't hugely nourishing, so a bullfinch has to shred hundreds of them to find enough of the tiny nutritious growing hearts to sustain itself. One bullfinch can strip a tree at a staggering pace, demolishing 10 to 30 buds per minute – that's half the buds on a pear tree in a single day.

Occasionally, the bullfinches' travels take them through gardens. They rarely land on a bird table but make a beeline for any seed feeders, especially if they're full of sunflower seeds, or shelled peanuts (surprisingly, their beaks are not strong enough to open peanut shells). Then they may explore the garden, always on the lookout for berries, seedheads and pods: early in the year they concentrate on cracking the pods of ornamental shrubs such as *Weigela*; come the summer, soft fruits are the main draw; and in autumn they feast on the seeds in *Delphinium*, *Antirrhinum* and *Campanula* pods and sunflower hearts.

Facts

Frequency

Increasingly scarce in woodland, hedgerows and orchards throughout Britain, numbers of bullfinches have fallen by nearly two thirds in the past 40 years. The causes of the population decline aren't fully understood, but those old bugbears – the over-zealous use of agricultural herbicides and the loss of its hedgerow haunts – are likely to have played their part. Significantly perhaps, the fluctuation in the bullfinch's fortunes parallels the decline and recovery of the sparrowhawk, a bird of prey that targets hedgerow birds.

Identification

The female has a jet-black skullcap and beak like her mate, but her brownish-pink underparts and olive-grey back are less brazen than his hot-pink breast and grey back. Both have a prominent white rump, black wings and tail.

Song

Short, faint, whistling *deu-deu* notes sound rather sad, like muffled pan pipes, and are often missed.

Nesting

The female weaves a twiggy platform of twigs with a shallow nest cup, lined with rootlets and hair, deep in a thick, thorny or evergreen cover.

Length

14-16cm (5-6in)

Conservation

Even admirers of the handsome bullfinch can be a bit 'not in my back yard' about encouraging it into their gardens when there are fruit trees and strawberry plants involved. But planting seedheaded and berry-bearing plants can help bullfinches find enough food during the winter and spring, and help save the local fruit trees. Countryside Stewardship Schemes promoting hedgerow- and field-margin development are a priority to turn around the fortunes of the bullfinch.

'The honours of his ebon poll
Were brighter than the sleekest mole;
His bosom of the hue
With which Aurora decks the skies'

From 'On the Death of Mrs Throckmorton's Bulfinch' by William Cowper

Carrion Crow
Corvus corone

To be frank, the crow is a bird that few actively want to entice into their gardens. Their black plumage, raucous and aggressive demeanour, and plundering ways rarely endear them to gardeners, who also despise their habits of eating pretty well anything that's available.

In yesteryear, superstitions took the loathing of crows even further. It was considered unlucky if a crow sat on your rooftop, for example, for it would suggest that there would soon be a death in the house.

Having said that, crows are surprisingly wary birds considering their size and strength.

They've learnt how to live alongside mankind, but they still keep a watchful eye on human movements, dropping into gardens more when backs are turned than in full view. As time goes by they develop in confidence, though, and may come to feeding tables, scattering smaller birds in the process.

There are two main ways of telling whether there's a rook or a crow sitting in your tree at the end of your garden. The first is the look of the bird – the rook's white beak, bare face and pointed head differentiate it from the crow. The second approach, although less reliable, is to see

Adult crows have brown eyes, but their young have blue eyes.

Facts

Frequency
The carrion crow is a common bird. It is found in virtually every environment in England and Wales, from city centres to coastal regions, woods and forests to moorlands. In northern Scotland and Ireland, however, its range peters out, and it is replaced by the hooded crow.

Identification
The crow is inky-black all over, and close up can only be confused with the raven, which is a much bigger and bulkier bird.

Song
A hoarse 'krark' is the main call, with a number of variations on the theme.

Nesting
Both male and female work on the nest, which is a large construction of sticks and twigs lined with bark and horsehair. Crows nest wherever they can – in tree forks, cliff edges, or even in the scaffolding of an electricity pylon.

Length
47cm (18.5in)

Did you know?

■ Young Crows are truly omnivorous birds, eating everything from insects and seeds to young birds, eggs and, of course, carrion.

■ Crows that live in coastal regions can often be seen feeding on shore crabs and mussels which they break into by dropping them from a height.

'The crow will tumble up and down
At the first sight of spring
And in old trees around the town
Brush winter from its wing.'
From 'Crows in Spring' by John Clare

if it's alone. As the old saying goes, if you see lots of crows together, then they're rooks. But if you only see one rook, then it's a crow.

The question is, which crow? Across England and Wales, the carrion crow holds court, but once you get into central and northern Scotland, as well as Wales, it is replaced by the hooded crow, with its distinctive grey nape, back and underparts. Until recently the hooded crow was thought to be a subspecies of the carrion crow, but a few years ago the two birds were split into separate species.

Chaffinch
Fringilla coelebs

For most of the year, chaffinches are such a familiar part of the garden scene, yet it is easy to overlook them as they hop modestly about the lawn under the bird table. But in the spring, the male chaffinch is a glory to behold in his breeding plumage: over the winter, the tips of his sombre autumn plumage get worn away to reveal a brilliant chestnut breast and cheeks topped by a slate-blue helmet, which make him one of the most strikingly handsome and colourful birds about.

In spring, you suddenly realise that the garden would be a much drearier place without a chaffinch or two in residence. The male chaffinch is not only one of the most handsome and colourful birds about, but he also has a delightful singing voice and is one of the keenest and most persistent singers in the garden choir. To make himself visible to prospective mates, the male perches on a prominent singing post, sticks out his chest and delivers his gleeful anthem with all the vigour and joy he can muster. Hearing his early practice sessions in February is a welcome reminder that it will soon be springtime again.

During the breeding season, chaffinches have to find plenty of caterpillars and other insects to feed their chicks in the nest. Otherwise, for the rest of the year, they are busy foraging for grains and seeds in gardens, woodland and hedgerows. Over the winter months, they are bird-table and seed-feeder regulars, although they prefer to hop around underneath, pecking up grains and seeds that have fallen to the ground, and crack them with their strong beaks. In the years when beech trees produce lots of beechmast in the autumn, you may not see so many chaffinches in your garden over the winter, as they flock to beech wood floors to extract the beechnuts from their tough seed cases.

Did you know?

■ The chaffinch features on one of the earliest registers of British birds ever recorded: a rare Anglo-Saxon document, dating from around 685, compiled by St Aldhelm of Malmesbury.

■ In summer, the male chaffinch's beak is grey-blue but fades to pale brown for the winter.

■ The rhythm of the chaffinch's song – *chip-chip-chip-chwee-chwee-tissi-chooeo* – has been compared to a bowler delivering a ball in cricket: the first six notes are the equivalent to his run-up and the final one represents the ball sailing through the air after it has left the bowler's hand. The entire song takes only two or three seconds to complete, but is repeated many times over by an enthusiastic male.

■ Years ago, the chaffinch used to be known as the 'bachelor finch', because the males and females coming from Scandinavia to spend the winter here often formed into large single-sex flocks. This habit was formalised in the bird's official scientific name, *Fringilla coelebs*: *fringilla* is 'finch' in Latin and *coelebs* means 'unmarried'.

Facts

Frequency

A recent survey recorded nearly two chaffinches per garden, making it the fifth most common bird in Britain. The winter population is boosted substantially by large flocks of migrants, mainly females, escaping bitterly cold weather in Scandinavia and northern Europe. The visitors spend most of their time feeding on farmland while small groups of residents frequent gardens, hedgerows and woodland. Nearly six million breed in Britain each summer and, if anything, numbers are increasing.

Identification

True to finch form, the male sports the jazzy colours while the female is dowdier, with dull olive-brown upperparts and pale dun underparts. Both have a conspicuous white shoulder patch, a white wing bar and white outer tail feathers, which help to identify them in flight.

Song

Enthusiastic and much repeated *chip-chip-chip-chwee-chwee-tissi-chooeo*; frequent metallic *pink-pink* contact call; *chip-chip* flight call.

Nesting

The female builds a beautifully tidy nest in the fork of a tree or in a thick shrub, painstakingly weaving moss, dry grass and fine rootlets into a neat cup which she lines with hair and feathers and decorates with lichen on the outside to make it blend into the background.

Length

14.5-16cm (6in)

Gardening for Birds

Bird-friendly gardening is an essential way of encouraging birds into your garden. Here, **Helen Bostock of the RHS** provides an overview of how to attract them to your garden, what plants to grow, and how to look after them once they've made your garden their home.

Providing natural food

During the summer nesting season, most birds, including those classed as seed eaters, will collect insects and other invertebrate animals to feed to their young. There is not much that gardeners can do to enhance the insect supply, since to do so might jeopardise the health of the garden plants. It should, however, be borne in mind that insects and other invertebrates are an essential part of many birds' diets, and their presence in gardens should be tolerated if no obvious harm is being caused to plants.

There are many garden plants that provide food in the form of berries or seeds. Much of this food becomes available in the late summer/autumn, when birds need to build up their fat reserves for the coming winter. The true bird gardener will not, of course, complain if the berries are stripped off within a few days of becoming ripe!

Cultivated plants

Berberis spp. (B)
Cotoneaster spp. (B)
Crab apples, *Malus* spp. (B)
Firethorn, *Pyracantha* spp. (B)
Sorbus spp. (B)
Currants, *Ribes* spp. (B)
Holly – female cultivars,
   Ilex spp. (B)
Privet, *Ligustrum ovalifolium* (B)
Daphne mezereum (B)
Honeysuckle, *Lonicera* spp. (B)
Some single flowered ornamental
   cherries, eg *Prunus avium*,
   P. Cerasus (B)
Some rose species, eg *Rosa
   rugosa*, *R. moyesii* (B)
Viburnum betulifolium (B)
Oregon grape, *Mahonia* spp. (B)
Photinia davidiana (B)
Thorns, *Crataegus* spp. (B)
Sunflower, *Helianthus annuus* (S)

Native plants

Blackberry, *Rubus fruticosus* (B)
Elderberry, *Sambucus nigra* (B)
Hawthorn, *Crataegus
   monogyna* (B)
Alder, *Alnus glutinosa* (S)
Birch, *Betula pendula* (S)
Holly – female plants of *Ilex
   aquifolium* (B)
Ivy, *Hedera helix* (B)
Yew, *Taxus baccata* (B)
Guelder rose, *Viburnum opulus* (B)
Wayfaring tree, *Viburnum
   lantana* (B)
Purging buckthorn, *Rhamnus
   catharticus* (B)
Alder buckthorn, syn. *Frangula
   alnus* (B)
Wild roses, eg *Rosa canina*,
   R. rubiginosa (B)
Mountain ash, *Sorbus
   aucuparia* (B)
Whitebeam, *Sorbus aria* (B)
Musk thistle, *Carduus nutans* (S)
Field scabious, *Knautia arvensis* (S)
Devil's bit scabious, *Succisa
   pratensis* (S)
Greater knapweed, *Centaurea
   scabiosa* (S)
Teasel, *Dipsacus fullonum* (S)

Seed-feeding birds can be catered for by delaying the cutting back of annual and herbaceous plants until late winter. The withered foliage will also provide hiding places for over-wintering insects and spiders, and so give insectivorous birds a feeding area.

Listed on the right are some cultivated and native plants that provide either berries (B) or seeds (S) for birds. Some wild plants that occur in gardens as weeds, such as groundsel, chickweed, dandelion, fat hen, thistles and nettles, are good providers of seeds but are too troublesome to be encouraged. Some of the plants included in the list will sometimes seed themselves freely, but they have the merit of being sufficiently attractive to earn a place in gardens, especially in semi-natural areas. Some of the native trees and shrubs are also available as named cultivars; these are also likely to be good for birds, especially where the cultivar has been selected for its superior berrying.

Handy hint

When making new garden features, such as walls and sheds, consider whether accommodation for birds can be included in the design. Gaps in the stone or brickwork will allow wrens, wagtails or spotted flycatchers to use them as nest sites. Pre-formed 'brick' nest boxes are even available for this very purpose. Leaving gaps in the top of shed doors ensures wrens and swallows have access to a dry dwelling. Avoid over tidiness in the shed itself – an old working boot or disused gardening hat is a traditional nest spot for resident robins.

Going native

Royal Horticultural Society

A fundamental part of responsible gardening involves planting native species, which in turn protect and ensure the survival of native ecosystems. Birds are far more likely to flock to a garden full of familiar insects and plants, than a garden full of exotic delights.

Here, **Helen Bostock of the RHS** focuses on wildflower gardens, which can consist entirely of native plants, replicating a natural plant community. Alternatively, a wildflower garden can be designed to create a certain effect, without being too purist about whether the plants are native or not, and there are many steps in between.

To be successful, the establishment of native wildflowers in gardens requires some thought about the mode of introduction and the management regime throughout the year.

When planning a wildflower area, assess the site, taking into account climate, soil type, drainage and the degree of shade and shelter.

The growing conditions largely determine the plants that will thrive and look natural in the setting. It is better to adapt the selection of plants to the site rather than the other way round.

Observing local wild flora will indicate which species to grow.

Handy hint

It is well known that thrushes use large stones as an anvil to break open snail shells. Provide for them by deliberately positioning suitable stones at the back of the flower bed or the base of a hedge. If thrushes frequent your garden, it will not be long before a cluster of broken snail shells will tell you that the birds are using them.

Handy Hint

Guard against creating hazards in your garden for birds. Loose netting thrown over ponds to keep out autumn leaves or anti-bird netting over brassicas and fruit cages is lethal if birds get caught up in it. By keeping the net taut and the edges tucked in, there should be fewer casualties.

Where have all the wild flowers gone?

- *In Britain, one in five wild flowers is currently threatened with extinction*

- *98% of our flower-rich meadows have been destroyed in the last 60 years*

- *Half of our ancient woodlands have been lost since 1945*

- *On average, each county loses a species every two years*

PLANTLIFE is the only UK charity dedicated to protecting wild plants. Our work ranges from recovery projects to save more than 100 threatened plants and fungi to the management of nature reserves where wild flowers can flourish.

PLANTLIFE
our plants *our* planet *our* future

JOIN US *and you can help protect wild plants and their habitats*

- -

SIGN UP TODAY *and enjoy the full benefits of membership*

Mr/Mrs/Miss/Ms/Title

Name:

Address:

Postcode:

E-mail:

Choose your own subscription level

☐ **£2** monthly ☐ **£3** monthly ☐ **£4** m

☐ **£5** monthly ☐ **£10** monthly ☐ **£20**

OR a yearly amount of

☐ **£24** ☐ **£30** ☐ **£40** ☐ **£50** ☐ **£100** ☐ **£**.......... Oth

Please return this form to: **Plantlife, FREEPOST LON 10717, SALISBURY SP1 1BR.** Thank you for your support

PAYMENT BY DIRECT DEBIT
DIRECT Debit

Name(s) of Account Holder(s)

Branch sort code: ☐☐ ☐☐ ☐☐ Account no: ☐☐☐☐☐☐☐☐

Name and full postal address of your Bank or Building Society
Banks and Building Societies may not accept direct debit instructions for some types of account

To: The Manager

_____ Bank or Building Society

Address

_____ Postcode

Originator's identification number: Reference number (for office use only)

☐7 ☐2 ☐5 ☐4 ☐5 ☐6   ☐☐☐☐☐☐☐☐

Instruction to your Bank or Building Society. Please pay Plantlife International direct debits from the account detailed on this instruction subject to the safeguards assured by the Direct Debit Guarantee. I understand that this instruction may remain with Plantlife International an if so, details will be passed electronically to my Bank/Building Society.

Signature(s)

Date

Data Protection: As a member you may like to receive information about the work of other reputable organisations. If you would prefer not to receive this information, please tick the box. ☐

Plantlife International - The Wild Plant Conservation Charity is a charitable company limited by guarantee. Registered Charity No: 1059559 Registered Company No: 3166339 BYG0 Registered in England. Head Office: 14 Rollestone Street, Salisbury, Wiltshire SP1 1DX Tel: 01722 342730

lawn-seeding practices. Sowing thinly at a rate of 5g/m² for grass seed mixes and 1g/m² for pure wildflower seed will suit most sites and allow the establishment of the slower-growing species.

To achieve even sowing, mix the seed with a carrier, such as dry silver sand or barley meal (available from pet shops), and sow at half the rate in two directions. Water thoroughly after sowing and if conditions remain dry. The best time to sow wildflower seed is in September. Some seeds will germinate almost immediately, but some may require low winter temperatures to break dormancy and initiate germination the following spring.

On heavy soils, which become waterlogged in winter, sowing in spring is advisable, as some seeds may rot in wet soil. If sown in spring, germination of some species may not occur until the second year.

Seeding new sites

Drifts of wildflowers can be created in the garden, but many gardeners want a traditional flowering meadow of grasses intermixed with wildflowers.

If this is the aim, it is important to avoid vigorous grasses, such as rye grass and cocksfoot. Proprietary seed mixtures are available, related to the type of soil and situation. A mix containing around 85% grasses and 15% wildflower seeds is suitable.

Prepare the seedbed as for a new lawn, removing stones and breaking up clods of soil. Allow six weeks for the soil to settle and for weed seeds to germinate. Spray or hoe these off before sowing and do not fertilise the soil in any way. Many wildflowers colonise poor soil and it will upset the balance and encourage excessive vigour in the grasses if extra nutrients are added.

On very fertile soils it may be an advantage to remove the

topsoil but, for anything other than the smallest area, this requires machinery. Subsoil sites will retain an open sward for several years, with bare patches of ground ideal for natural colonisation. An alternative approach, on soils other than clays and those with high organic matter, may be to put the land down to mustard for a season to reduce fertility, removing the crop at flowering time.

Once the ground is prepared, seed can be broadcast and raked in, following standard

Handy hint

Sadly, many birds can stun themselves or even die through flying into shed, house, greenhouse or conservatory windows. Patio doors are a particular problem due to their large size. One way to avoid this happening is to stick silhouettes of birds, especially birds of prey, onto the glass – black for the window where there is another window at the other end and birds might think they can fly through, and red for a window or glass door that is reflective. These can be made from sticky paper or bought in garden centres.

Top 10 wildlife plants

For the full wildlife effect, you'll want your garden alive with butterflies and bees, as well as birds. **Joyce Millar of Northern Ireland's Ladybird Nursery** reveals 10 of the best plants to grow to attract birds and other wildlife.

Growing plants for wildlife gives such a special pleasure. Not only do you get to enjoy the plants, you also get to enjoy the countless creatures that share these plants with you. I think even the most hardened of us must feel uplifted when we see a butterfly flitting haphazardly around the garden.

It is difficult to define what makes gardening for wildlife so satisfying, but I think it has to do with a quality that can be difficult to achieve otherwise – the relaxing atmosphere that is created from the drone of bees or the beautiful songs of birds.

There are a great many plants that have value for wildlife – not just British natives. It is best to select plants that suit your conditions, rather than struggling with plants that have to be constantly cajoled just to survive.

Limiting chemical use is of great benefit. For the past few years, we have not used pesticides, slug pellets and so on in our garden. We rely on our birds, frogs and ladybirds to do that job for us – it's much easier and more rewarding.

Having something in flower in all seasons is important as a food source. Also, when trying to attract specific species, it is more effective to plant in groups – a group of plants to attract butterflies should be planted in one sheltered spot, rather than having individual plants being dotted around.

Increasingly, there are a great many sources of information on what to plant for wildlife, but a lot of fun can be had by just observing what works in your own garden. It has been difficult to narrow my own personal favourites down to 10, but here goes...

Handy hint

Try to position several feeding stations around the garden and move them around at intervals to prevent a build up of disease in the underlying soil and to allow turf to recover. If germinating seed beneath the feeder is a problem, pave over the area just below the feeder. Using de-husked seed and ensuring a surplus of food is not left on the bird table will also help. Sometimes it is quite fun to see what unusual plants grow from the seed – buckwheat, normal wheat, sunflowers, ragweed (*Ambrosia artemisifolia*) – but if problematic, these can be hoed off or knocked back with a glyphosate-base weedkiller.

Verbena bonariensis (below)

For a well-drained sunny spot, this is an exceptionally good plant. It will flower from early summer right through into November and possibly beyond, depending on the severity of the winter. It is also highly versatile – garden designers describe it as 'transparent' so, although it is tall, at 1-2m (35-79in), it can be planted even at the front of borders. Its lilac-purple blooms blend well with many other colours and its wiry stems make for a wind-resistant plant.

It is a butterfly and bee 'magnet', attracting many different species. What's more, by leaving the seedheads in winter many birds will be attracted – we have had siskin, redpoll and lots of finch and tits feeding on ours. Its only real downfall, for which it can be easily forgiven, is that it will freely self-seed when happy.

Centranthus ruber (valerian, above)

Grows best in poor, alkaline, well-drained soils. This is another good-value plant because of its long and vivid flowering season. With its clusters of honey-scented flowers, it is a rich source of nectar for many insects, including butterflies, bees and moths. Last year, we were thrilled to see hummingbird moths feeding on our valerian – definitely an unusual species here in Northern Ireland. It is a prolific self-seeder, so take care with location. We plant ours in pots (it doesn't need a lot of watering) and put it on a concrete patio. It will self-seed in the cracks, but these can be weeded out easily.

Eupatorium purpureum (Joe Pye weed, above)

Best in moister conditions, this plant is useful to the wildlife gardener as it is a butterfly plant that will tolerate some shade. It is a big bold majestic plant, so it needs a good space, preferably at the back of a border. The first time I saw this plant it became a 'must-have' plant instantly, as it was absolutely alive with butterflies. We see peacock, painted lady and tortoiseshell butterflies in our garden.

Echium vulgare (vipers bugloss)

For well-drained sunny spot, this charming biennial should definitely not be dismissed because it is not a perennial. We grow it in a gravel area and I am delighted each year by its reappearance from self-seeding. It is useful for low-level interest, never getting much higher than 30cm (12in) in our garden. Bees love it and with flowers of the most enchanting shades of blue, they are not alone.

Origanum vulgare (oregano)

Better known for its use in cooking, oregano is a really easy garden plant. Preferring well-drained soil in full sun, bees and butterflies love it and it provides food for seed-eating birds in winter. There are many varieties, some with attractive golden and variegated leaves – we often use them as foliage plants in pots. It's an ideal plant for a small garden because of its multiple uses.

Miscanthus sinensis

Grasses are an integral part of a naturalistic habitat. We find that they provide shelter for our wildlife, particularly over winter. Like many 'spring watchers', I look forward to the first sightings of ladybirds. I always go to our large clump of Miscanthus on the first sunny days, in the hope of seeing the ladybirds emerging from the clump as if they have just come out to sunbathe.

Many varieties of Miscanthus are available. They are mostly large plants, tolerant of a wide range of conditions, but are best in well-drained moist soil in full sun.

Handy hint

Take care with opening upstairs windows in the summer. House martins may be nesting under the eaves, or there may be nests in dense climbers, such as Virginia creeper or ivy. If these climbers need clipping back, again be very cautious or preferably delay doing this until the birds have left the nest.

Dipsacus fullonum (teasel)

For any reasonably fertile soil in full sun or partial shade, teasels are grown to attract goldfinches and other seed-eating birds to the garden. Less well known is its ability to attract nectar-loving insects to the striking and unusual flowering heads. With its architectural shape, it can be tempting to give it a prominent position in the garden, but because of its prolific self-seeding habit it is best sited in a wilder area. Children love the cups of water that collect insect casualties at the base of leaves, but care needs to be taken as the backs of the leaves have sharp thorny spines.

Ajuga reptans 'Burgundy glow' (bugle)

Best in moist well-drained soil in sun or partial shade, *Ajuga* is a really useful evergreen carpeting plant. We use it as ground cover for areas that we don't want to interplant, such as tricky areas below trees. We are rewarded early each spring with magnificent blocks of blue flowers, absolutely buzzing with bees. Although quite a common plant, it is very worthwhile to grow.

Handy hint

In the spring, put out nesting material. Sheep's wool is ideal as it naturally repels the wet. Simply hang this up in a loose net bag. Try this also with pet hair, collected when grooming cats or dogs. Many birds will use moss to line their nests. By hanging up old moss rings, left over from Christmas wreaths, birds can tease it out if they so wish.

Echinacea purpurea (coneflower)

Best in a deep well-drained humus-rich soil in full sun. Well-grown, *Echinacea* have substantial garden presence, with their large pink daisy flowers and each flower with its prominent central cone. Later flowering than most perennials, they provide a valuable source of nectar for insects preparing for hibernation. We plant ours close to other butterfly- attracting plants, such as verbena and asters, and look forward to the droves of butterflies that inevitably appear. At over 1m (39in), these are best placed in the middle or back of a border.

Rudbeckia fulgida var. *sullivantii* 'Goldsturm'

This is a useful late-season butterfly and bee plant for well-drained soils in full sun or partial shade. As summer draws to a close and colours begin to wash out, I love this plant because it can be relied upon to be at its bright, vivid best. Bright golden flowers, with a prominent dark cone are freely produced until the first frosts. It is said that the petals have ultraviolet markings to guide insects to the central cone. We always leave the seedheads over winter as food for birds. A favourite combination is with a purple *Cotinus* shrub as a backdrop.

About Ladybird Nursery

I started the nursery in 2005 with the aim of combining two passions – gardening and wildlife. Not so long ago, wildlife gardening was often considered to be a bit of a joke. I wanted to show that you could have a beautiful garden, with plenty of wildlife and that it was quite easy to achieve. As well as our interest in plants for wildlife, we stock an eclectic mix of herbaceous plants, grasses and choice annuals. The nursery is in our ¾-acre garden, which we open by appointment and on special days for charities. **Ladybird Garden Nursery, 32 Ballykeigle Road, off Moss Road, Near Ballygowan, County Down, N Ireland BT23 5SD www.ladybirdgarden.co.uk**

Chiffchaff
Phylloscopus collybita

Here's a quick quiz question. What do the following birds have in common: turtle dove, cuckoo, chiffchaff? The answer is they're all onomatopoeic, their names mimicking their calls. In the case of the chiffchaff this comes as something of a relief, because without its distinctive song it's difficult to differentiate from its close cousin, the willow warbler.

In fact, the two birds are so similar that many birdwatchers lump them together as 'willowchiffs', but while they're in spring- and summer-song mode, salvation is at hand. The chiffchaff lives up to its name, its song being a random and staccato sequence of *chiffs* and lower-pitched *chaffs*, while the willow warbler sings a gentle cascade of fluid notes.

Early-March chiffchaff arrivals from the Mediterranean or Africa are one of the first signs that spring is here, as the little birds with their olive-brown complexion flit about garden tree branches and shrubbery on the constant lookout for insects. Often hanging out with families of similarly sized blue and great tits, they're just what the gardener ordered. The aphid that escapes the chiffchaff's attention is a well-hidden aphid indeed, as the bird is one of the more thorough bug hunters. Agile, energetic and dedicated, it checks leaves top and bottom for tiny insects, sometimes even snatching them from the air in mid-flight.

Times are changing though, and the warmer British winters are encouraging more and more chiffchaffs to drop in at this time of year. If you spot a twitchy little warbler flicking across the end of your garden next winter (and it doesn't have a black or brown cap, which would make it, unsurprisingly, a blackcap – another over-winterer), then it's more than likely to be a chiffchaff. Some birds decide not to go back to Africa, others drop in from Scandinavia to spend their winter here instead.

Facts

Frequency
Numbers fell away in the 1970s after a series of colder African winters, but have been climbing back up ever since. One of our commonest warblers, the chiffchaff is quite a frequent garden visitor, particularly if you live close to woodland.

Identification
A neat, busy little warbler. Look out for its dark eyestripe and subtle light streak above the eye. To differentiate from the willow warbler, song is your best bet (see left), but the chiffchaff also has darker legs and slightly duller plumage. When on the move, the chiffchaff wags its tail, whereas the willow warbler presents more of a flick. The chiffchaff has more of a preference for coniferous trees, too. The over-wintering birds are slightly greyer than summer visitors, and ornithologists are still trying to work out exactly which races they all are.

Song
The chiffchaff belts out its own name. The autumnal call is a thin, whistling *hooeet*.

Nesting
Surprisingly close to the ground, often in tall grass or low bushes. Dead leaves and twigs form the nest's base, while feathers are used to create a thick lining.

Length
11cm (4in)

Did you know?

■ Only the female chiffchaff feeds its young.

■ The bird's Latin name, *Phylloscopus*, means 'leaf explorer' – a perfect description of the chiffchaff's feeding habits.

■ The pale 'eyebrow' of the chiffchaff, and other birds, is called a supercilium.

Coal Tit
Parus ater

Of the three kinds of true tit that commonly live in gardens, the coal tit is the smallest and probably least well known. Imagine a rather bleached-out mini great tit tumbling about with all the energetic charm and agility of a blue tit and you've got the coal tit. Then look for a distinguishing white patch on the back of the neck to confirm its identity.

All three were originally woodland birds, feeding mainly on insects in trees and nesting in tree holes. To avoid coming off third best against its bigger rivals in the contest for food and nest holes, the coal tit specialises in living in coniferous woodland and trees. Its fine beak is ideal for probing between needles and delving into cracks in the bark for insects or extracting seeds from open pine cones. It also has larger

Did you know?

■ Coal tits wedge surplus nuts and seeds into slits in bark, among pine needles or in patches of moss and lichen, all places they would normally look for food. Usually a coal tit eats its stores within a few days, as long as the great tits don't get there first and it can remember where it stashed them. If a sunflower starts growing in an odd place around the garden, you could blame an absent-minded coal tit that forgot where it left the seed.

■ The coal tit is the tiniest tit in Europe and one of the smallest birds in Britain.

■ Blue tits permitting, coal tits will nest in nest boxes with small entrance holes in the garden. You can help by positioning a box low down, near the ground, because blue tits prefer to nest higher up.

feet and longer claws than other tits, which give it a good grip on the bark and slippery needles. In Poland, the most common name for the coal tit translates as the pine tit.

Generally, coal tits are attracted to gardens where there are pines, firs, cedars, larches and yews growing. Even so, you're unlikely to see much of them during the summer, when they spend most of their time in the upper branches, searching for insects – aphids, moths, caterpillars, weevils, flies and bugs – and raising their families. But come the winter, coal tits become regular visitors to peanut feeders and lightning raiders of the bird table to snatch seeds and suet, which they carry off to eat in peace elsewhere: if they hang about too long, they're likely to be mugged by great and blue tits. In mild winters, coal tits may gang up with other tits and roam through the local woodland, dividing up the spoils between them. Unlike other tits, the coal tit is happy to feed on the ground, especially under beech trees that have dropped plenty of beech mast.

'Hopped on the bough, then, darting low,
Prints his small impress on the snow,
Shows feats of his gymnastic play,
Head downward, clinging to the spray'

From 'The Titmouse' by Ralph Waldo Emerson

Facts

Frequency
Common and widely distributed; over 650,000 pairs breed in Britain each summer.

Identification
Its pale colouring and lack of any bright yellow or blue plus the definitive patch of white feathers on the back of the neck mark it out from the other black-capped tits; both male and female have a black cap, throat and bib (it's hard to see that his bib is slightly larger and more triangular than hers), white patches on the cheeks, olive-grey upperparts, double white wingbars, pale-buff breast deepening to brown on the flanks, belly and rump; juveniles have a yellower tinge.

Song
Male sings *weecho-weecho* almost all year round, especially between January and June; frequently uses a plaintive high-pitched *tsuee-tsuee* call, which may be mistaken for the call of another tiny conifer-dwelling bird, the goldcrest.

Nesting
Prefers to build in empty woodpecker holes or natural tree holes, low down on a dead or decaying stump, but may have to use crevices in stone walls, among tree roots or empty mouse holes; female assembles a cup of moss and dry leaves woven together with spiders' webs and lined with wool or rabbit and mouse fur, and the occasional feather.

Length
10-11cm (4in)

Collared Dove
Streptopelia decaocto

Sitting on the patio or lying in bed in the morning, listening to the neighbourhood collared doves *koo-kooOO-kuk* cooing away, it's hard to believe that they've been living in Britain for only 50 years or so. Now, collared doves are an ever-present part of the garden scenery and soundtrack. In most towns and villages, you're probably never more than about a mile away from a collared dove or two.

Collared doves are almost always seen in pairs or small flocks. You'll often see them perched together on telephone wires, chimneys, TV aerials and roof ridges. Apparently, partnerships last a lifetime; when one flies off, the other usually flaps after it with a gull-like *kwerr* call, and they frequently indulge in intimate billing, cooing and mutual preening to reinforce the pair bond. When a male wants to show off to a female, he claps his wings as he takes off from the perch, rising steeply into the air. Then he

Did you know?

■ From one pair in Norfolk in 1955, there were more than 400,000 collared doves in Britain by 2000. Originating in India, they moved west in the 16th century, but settled down when they reached Turkey. In 1912, collared doves started advancing across Europe, reaching France and Belgium in the early 1950s. From there it was a short flight across the North Sea to East Anglia.

■ In August 1952, a strange dove was seen feeding in a chicken run, near Manton in Lincolnshire. Reg May, a local postman and keen ornithologist, saw it as the first male collared dove to set foot on British soil. A pioneering pair bred near Cromer in Norfolk in 1955 and their descendants have been going strong ever since, reaching Surrey in 1956 and Lewis off the west coast of Scotland in 1960. By 1964 there were already roughly 300 breeding pairs; that rose to 100,000 by the early 1980s. Today,

there are approximately 200,000 pairs all over Britain.

■ Collared doves spend a lot of time preening. They have special, small, powder-down feathers – which grow continuously and keep crumbling to produce a white dust which covers the whole bird and keeps the other feathers clean and waterproof. You may see them sitting in a summer shower, stretching their wings and freshening up before rearranging their feathers with their beaks.

■ When a collared dove visits a garden bird bath, it can suck the water up through its beak like a straw. Other birds have to scoop-tilt their heads back to let the water drain down their throats.

■ The collared dove has now set its sights on conquering North America. It first landed in Florida in the early 1980s and is known as the beige starling, a back-handed tribute to its local abundance and non-stop cooing.

glides down again on fanned wings and tail to land back beside her and bows deeply.

Many collared doves base themselves in gardens where everything they need is available: year-round cover for roosting and nesting, supplied by evergreens and conifers (although other environments work well, too) plus well-stocked feeders and bird tables. (It may be no coincidence that the rise of the collared dove paralleled the increase in planting of, leylandii, for hedging and in garden feeding.) Mostly, collared doves scavenge seeds and grains on the ground, although they also take grain that's ripening in fields, and berries in trees, hedges, shrubs and creepers. They're becoming more adept at landing on hanging bird feeders fitted with trays. One reason collared doves do so well is that they can gulp down whole seeds faster than finches and sparrows, which have to crack the seed cases before swallowing them.

Another key to the collared dove's great success is its breeding capacity. In theory, they could breed throughout the year, because they produce a nourishing crop-milk to feed their chicks, rather than relying on a seasonal food supply, such as caterpillars. But given the unpredictability of the British climate, a pair of collared doves usually raises just four sets of twins between March and October. As parents, they're valiant in defending their eggs and chicks from nest robbers – chasing off magpies, which other garden birds would never dare do.

When the collared dove showed signs of prospering here, its progress was monitored nervously to see what impact the invader would have on the native birdlife. It became clear that the collared dove had filled a gap in the market place without evicting a native resident or regular winter or summer visitor. Until the collared dove turned up, there was no medium-sized, grain-eating, tree-nesting bird in Britain.

Facts

Frequency
Ranks among the most abundant and wide-ranging of all garden birds. One of the larger birds to visit bird tables regularly.

Identification
The subtle pastel blend of pinkish-sandy browns and greys in a collared dove's plumage gives it a delicate beauty. It takes its name from the black half-collar finely outlined in white around the nape of the neck of both male and female; juveniles develop a collar after their first moult.

Song
In spring and summer, the male coos almost continually, especially early in the morning. His soft, throaty *koo-kooOO-kuk* is composed of three notes – the first two are high, with stress on the drawn-out middle note, the third lower.

Nesting
The male starts gathering twigs and brings them back to the female at the nest site, usually in a conifer or evergreen tree, where she arranges each one into a flimsy platform around her.

Length
31-34cm (12-13in)

Conservation
Up and coming – if anything, the consequences of its phenomenal success, particularly its droppings and monotonous cooing, are likely to make it unpopular in some quarters.

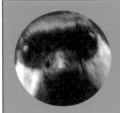

David North, Norfolk Wildlife Trust, advises:

■ Any list of Britain's commonest garden birds will now include the collared dove. Yet the arrival of Britain's first breeding pair near Cromer in Norfolk in 1955, is just part of an amazing avian success story. Before 1930, collared doves had only a toehold in Europe, with breeding confined to the Balkans. This species, for reasons still not fully understood, then began an amazing expansion westwards. Collared doves have now successfully conquered Europe in a bloodless coup that seems to have allowed them to pioneer a niche occupied by no other bird. RSPB's long-running annual Big Garden Birdwatch has monitored their numbers in gardens between 1979 and 2006 and shown an amazing 416% increase.

■ Part of their success reflects their unusual breeding behaviour, with young having been recorded in nests in every month of the year. The nest itself is little more than a handful of twigs balanced precariously in the fork of a branch, so flimsy that the eggs can usually clearly be seen when viewed from below! Clutches are small – two eggs is normal – but with three to six broods a year their reproductive capacity is phenomenal.

■ If you want to attract collared doves to your garden then provide grain using a ground feeder. They will visit bird tables, but are happiest on the ground and adore wheat, barley and mixed seeds, though will readily take bread.

■ Familiarity can breed contempt, or at least stop us seeing the beauty of a bird. The collared dove is an exquisite pinky-grey with a neat black half-collar around the neck. In flight, the white-tipped tail is especially noticeable. The *coola-coooo-cooo* song can be monotonous, but is now heard in gardens from suburbia to the countryside. Wherever people make gardens and feed birds, the collared dove is likely to be there, a fact which hasn't gone unnoticed by passing sparrowhawks, which now count the collared dove among their favourite prey.

Crossbill
Loxia curvirostra

Sometimes a bird's name says it all, and the crossbill is one of the best examples of perfect naming. Equipped with its unique bill in which both mandibles are elongated and hook past each other, the crossbill has the perfect inbuilt tool for prising open pine cones to get at the tasty seeds trapped inside. To keep close to their food source, they breed exclusively in conifers, particularly pine trees. If you don't have any conifers in your garden, you won't see any crossbills, but even a single tree might briefly attract at least one of these passing finches.

Bright red plumage for the male, and olive-green for the female, along with those distinctive bills, should make the crossbill extremely easy to identify. There are two drawbacks, though. The first is that the bird is very shy, and tends to trawl for its meals in the very tops of trees, making it hard even to see in the first place.

The second is that, where we once thought we had just one species of crossbill in this country, we now know we have four. Parrot and two-barred crossbills are occasional visitors to these shores from the Continent, but the Scottish crossbill, once thought of as a subspecies, has recently been verified as a species in its own right. What's difficult about this is that the common crossbill and the Scottish crossbill are difficult to tell apart (although the latter is a little bulkier and is more usually found in Scots pine). There's an upside, though. The Scottish crossbill is the only endemic British bird species, found nowhere else in the world.

Outside of Scotland, though, a crossbill can only be the common crossbill. If you're wondering if you've got any in your area, the best clue can be found on the ground. Mangled, twisted cones that have been dropped from on high are the best indicator that the finch with the overgrown bill is in town.

A nest of young crossbills can consume up to 85,000 pine seeds before fledging.

Did you know?

■ Young crossbills have bills that remain straight until they are fully fledged, and even for a few weeks beyond that point, to allow them to open their gapes wide enough to be able to accept the food their parents bring them.

■ The crossbill is one of the few birds that can be found on both sides of the Atlantic. In addition to its American presence, it can also be found in the Middle East and south-east Asia.

■ Because of its bill, the crossbill is incapable of picking up seeds from the ground, although it does come to the woodland or even garden floor to drink water.

Facts

Frequency
Only a few thousand pairs breed in the British Isles each year, but they are fairly well distributed. Main centres of population include the New Forest, Norfolk and several parts of north-east England. Scotland, however, is where they are most common.

Identification
Chunky red finches with crossed bills are hard to confuse with anything else – although it has been done. The bird's stocky nature, red or green plumage and hooked bills have led many birdwatchers to think that they've stumbled across an escaped parrot. The way in which the bird moves sideways along a branch, and uses its bill to manoeuvre itself into position, all serves to enhance this impression.

Song
High-pitched, thin twittering from the treetops might give away a crossbill's presence, as can its surprisingly vehement *chip-chip* flight call.

Nesting
The crossbill builds a cup of twigs, moss and grass high up in conifers. Unusually, these birds may nest at any time of year, depending upon the abundance of food.

Length
16.5cm (6.5in)

'In the groves of pine it singeth Songs, like legends, strange to hear.'

From *The Legend of the Crossbill* by Henry Wadsworth Longfellow

75

Dunnock
Prunella modularis

For most of the time, dunnocks keep a low profile in the garden, hugging the fringes of a lawn or creeping under a hedge or shrubbery while hunting for insects, spiders, small earthworms and tiny snails in the grass, leaf litter and undergrowth. Their brown and grey plumage blends into the background so well, only the constant twitching of their wings and tail gives them away at times.

As a dunnock shuffles along in the shadows, its whole body language suggests timidity. But the lengthening, warmer days of spring bring out the exhibitionist in male dunnocks and suddenly you see them in a fresh light. There they are, out in the sunshine, competing for the best territories and females by skipping along the tops of hedges, waving their wings about in a flamboyant, slow-motion version of their usual wing-flicking.

And then, when it comes to breeding, the supposedly modest dunnocks behave quite outrageously! They have complicated sex lives; unusually, both the male and the female are frequently unfaithful to each other. In years when food is plentiful, the male may mate with two or three females, who build their nests on his patch and tend their broods largely on their own. When times are tough, the female may seduce another male to help her bring up her partner's family. To give the second male

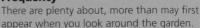

an incentive to stick around, she lets him mate with her as well.

Male dunnocks, evidently, know better than to trust their partners as they've developed an anti-cuckolding ploy to ensure that they don't waste time and effort rearing another male's offspring. When an excited female is prepared to mate, she crouches in front of her partner and lifts her tail. But before jumping on her, the male pecks at her behind, stimulating her to expel any sperm from an earlier coupling. In spite of his precautions, DNA paternity tests on dunnock chicks from the same nest often reveal that they have been fathered by different birds.

In the summer, when hedgerows and thickets are teeming with insects and the like, dunnocks may really disappear from the garden. There's no need to worry; they haven't come to any harm or gone for good – they'll be back in the autumn. As one of the few insect-eaters that dares to stay in this country all year round, the dunnock needs to augment its winter diet with seeds, often from ground-hugging weeds such as chickweed and plantain, plus any crumbs that get knocked off the bird table.

Facts

Frequency
There are plenty about, more than may first appear when you look around the garden.

Identification
Male and female look alike, with a tortoiseshell-like pattern of rich chestnut-brown streaked with black over the back and wing, and ash-grey underparts.

Song
The dunnock is surprisingly vocal; the male sings his high-pitched melody for much of the year, with extra fervour in the breeding season; both male and female constantly exchange plaintive *pseep-pseep* contact calls while foraging and issue thin *sissisiss* flight calls on take-off.

Nesting
The female builds a neat cup of dried grass and moss woven on to a base of fine twigs, lined with hair or wool low down in the thickest part of a hedge or bush, or among brambles.

Length
14-15cm (6in)

Did you know?

■ The female cuckoo is notorious for shirking her nest-building and chick-rearing duties by dumping her eggs in the nests of other birds. Over southern England, the poor old dunnock is the cuckoo's favourite foster parent. Although the grey egg is the odd one out in a clutch of sky-blue dunnock eggs, the dunnock incubates it along with the rest. Once the cuckoo chick hatches, the dunnock can do nothing to stop the interloper from heaving its eggs or chicks out of the nest. Again, the dunnock simply gets on with raising the cuckoo chick, and foregoes its chance of rearing a brood of its own.

■ Many older people may still call the dunnock by its former name, the hedge sparrow, which dates back to the 16th century. But there were always objections, as the bird bore only a passing likeness to a sparrow and wasn't related to it at all. After much debate, the hedge sparrow reverted to being the dunnock, a melding of 'dunn-cock', meaning grey-brown bird.

Fine Dining al Fresco

For three generations Walter Harrison's have been perfecting the art of fine dining for garden birds, and our new Garden Friends range is the tastiest ever to leave our mill.

To delight your garden diners ask for it by name. If your usual retailer can't oblige you, phone 0115 9332 901 or visit our website at www.walterharrisons.com for details of your nearest stockist.

Proudly Supporting

Philip Precey, Derbyshire Wildlife Trust, advises:

■ An inconspicuous bird, the dunnock can often be seen flitting about at the edge of the lawn, skulking around the base of the shrubbery, always on the move. Look closer and you may notice that they often travel in pairs, one following a step or two behind the other. It sometimes looks like the leading bird, the female, is trying to slip away from the other, flicking off sideways to shake its 'tail'. That's exactly what she's doing; if you look closer, you'll see a third bird – a second male – following to the side, some way back. In the dunnock world, adultery is rife, with the alpha female trying to give her mate the slip and get some quality time alone with the beta male. Just before you feel sorry for her cuckolded husband – he's not averse to nipping off to be with neighbouring females either! They're the most hot-blooded birds in the suburbs!

■ Dunnocks are insectivores, and rarely visit the bird table. Instead, make your garden as dunnock-friendly as possible. As woodland edge birds, they particularly like spending time in the interface between shrubbery and open lawn, where much of the female's flirting takes place, while the bushes and shrubs provide suitable sites for nesting. If you don't already have them, try to introduce some of the features that are more likely to attract as many insects and invertebrates as possible; log piles, rockeries, native planting, a compost heap or a pond will all increase the invertebrate diversity of your garden, and keep the dunnocks well fed.

Feral Pigeon
Columba livia

Woodpigeons, collared doves, stock doves, maybe even turtle doves – all these members of the pigeon family are likely or possible visitors to your garden, and each of them can be easily described and identified. The feral pigeon, however, is a different matter all together.

There are great ones, small ones, lean ones, brawny ones, brown ones, black ones, grey ones, tawny ones. Like the rats in *The Pied Piper of Hamelin* (and to many people these birds have a similar vermin status), feral pigeons come in all sorts of colours, shapes and sizes. The extraordinary variety stems from the fact that

they're descended from escaped domesticated pigeons that have been bred for centuries, like dogs and cats, for different purposes and to create different looks. In turn, domesticated pigeons have been bred from the rock dove, the bird of rocky coastal regions which some feral pigeons still resemble.

At times, they really do seem to be everywhere, and they're not the most popular of town garden visitors, munching away as they do on any available vegetable, fruit or seed and, to most pairs of eyes, offering little in return in the way of charm.

Feral pigeons are not fussy eaters – whereas humans have around 10,000 taste buds, pigeons only have 37.

Facts

Frequency
Common, abundant and never properly counted, feral pigeons occur wherever several humans gather.

Identification
Those that exhibit the strains of the rock dove are grey with slightly iridescent neck feathers that flash pink and green in sunlight. Two black bars on the wings help to mark them out from stock doves. For the rest, virtually anything goes. As pigeons have been bred for centuries by fanciers who try to develop new colours and styles, the progeny of those that escape are extremely varied.

Song
A soft *coo* repeated over and over. Some birds have harsher voices than others, though.

Nesting
These birds tend to nest on buildings, which replicates their ancestors' behaviour on cliff-tops.

Length
Approximately 29-33cm (11.5-13in)

Did you know?

■ Wild rock doves feed on seed, but their feral descendants will eat almost anything, including bread, meat and fish and chips left discarded in the street.

■ Although the feral pigeon is widespread, its wild cousin can only be found on the west coasts of Scotland and Ireland.

**'The pigeon with its breast of many hues
That spangles to the sun turns round and round
About his timid sidling mate and coos'**

From 'The Summer Gone' by John Clare

   Their courtship behaviour is always enjoyable to watch, though. The male follows his intended around with bowed head and fanned tail, wooing her with his cooing. A little two-step then takes place, in which the female holds the male's beak, and then shortly afterwards, if all has gone well, mating occurs.

   If you do have pigeons frequenting your garden, and you're not too keen about the idea, there is one thing to bear in mind: a pigeon is a very fine catch for a hungry sparrowhawk, and as a result your songbirds are more likely to be left alone.

Fieldfare
Turdus pilaris

The harsher the winter weather, the more likely you are to have fieldfares in your garden. These visiting thrushes from Scandinavia stay in Britain for longer than most migrants – arriving in October and leaving again as late as May – and they tend to prefer open fields in which to forage. When times are rough, however, and snow has fallen, they will follow redwings into even the most urban of gardens in a desperate hunt for berries or fallen fruit. *Pyracantha* or hawthorn are among your best bets for encouraging them – basically, the redder the berry the better.

The fieldfare is a somewhat sophisticated bird. Larger than Britain's other thrushes, its pencil-grey head and rich chestnut back lend it a somewhat debonair look – an image that is immediately shattered as soon as the birds take flight. On the wing it barks out a rather manic-sounding series of grating *chak*s as it thrusts itself along with a few rapid wing beats, then closes its wings and glides like a missile.

Gregarious creatures, fieldfares hang around together in groups of at least 30 birds, and sometimes as many as 200. You can often see them mixing with other members of the thrush family as they rove around fields foraging for

'The fieldfare chatter in the whistling thorn
And for the awe round fields and closen rove'

From 'Emmonsails Heath in Winter' by John Clare

Facts

Frequency
Apart from the northern reaches of Scotland, fieldfares visit the whole of Britain and Ireland during winter. About a million of them migrate here each year, and a tiny handful, perhaps a couple of dozen, stay over to breed.

Identification
A large thrush with a grey head, dark chestnut back and rump, and specked breast. The beak is yellow like the blackbird's, with the difference being that the fieldfare's has a black tip.

Song
During the breeding season, an apparently random collection of whistles, chuckles and high squeaks. In flight, a harsh *chak-chak-chak*.

Nesting
One or two broods reared in a nest of moss, twigs and grass, held together with mud and lined with grass.

Length
25.5cm (10in)

Did you know?

■ Breeding fieldfares in Scandinavia tend to choose woodlands where merlins patrol. The fieldfare is too big for the small falcon to bother with, but the merlin's predatory habits keep the fieldfares safer.

■ Fieldfares can live for up to 10 years.

■ In particularly hard winters, fieldfares have been known to invade the yards of greengrocers in the hunt for fruit.

The white feathers under the wing are useful for identifying fieldfares in flight.

worms, slugs, spiders and insects. These birds truly work as a team, covering the ground like sappers clearing a minefield, leaving not even a square centimetre unexamined.

As the day draws to a close, they fly up en masse to their chosen roost sites, where once again they appear to be birds with military training. The next time you see fieldfares settling down for the night, take a good look – they'll all be facing the same way, like an army viewing their distant target. What they are all actually doing is making sure they're facing the wind, in case they're disturbed and have to make a quick getaway.

> **'And in the little thickets where a sleeper**
> **For ever might lie lost, the nettle-creeper**
> **And garden warbler sang unceasingly'**
>
> From 'Haymaking' by Edward Thomas

It's sometimes said that the most distinguishing feature of the garden warbler is that it has no features at all. That seems rather cruel. What's even crueller is that some birdwatchers like to put a 'g' at the end of the bird's scientific name. The poor old garden warbler – it really is an overlooked bird.

Another reason for this is that it's not particularly well named. The bird does sometimes appear in gardens, but very rarely, as it's a secretive soul, preferring to hide itself away in deciduous and mixed woodland, rarely venturing further than the woodland edges or rides. If you have a mature garden alongside a wood, you might be lucky enough to have one pop in on its forage for insect life. In the autumn, though, like some other warblers, the bird turns to berries and fruit.

A summer visitor, the garden warbler most closely resembles a slightly darker blackcap without the cap. In fact, its song is very similar to its cousin's, too, although with practice you can separate the two. Listen carefully and you can hear the garden warbler's slightly mellower voice, with longer phrases. It's a gentle song, slightly wistful yet bouncy. It has been likened to the sound of a bubbling brook.

Sadly, the garden warbler is in slight decline, like several of its fellow trans-Saharan migrants. There were once more than 200,000 breeding pairs, but in the last few years this figure has been slowly dropping, due to the less hospitable climate in Africa where it spends the winter.

Facts

Frequency
The garden warbler starts to arrive in late April or May, and they've all gone by September. You can find it across most of Britain, except the north of Scotland, and in central Ireland. In August and September continental migrants can appear along the south and east coasts of England.

Identification
Brownish-grey above and pale below. That's about it.

Song
A pleasant chattering song, similar to the blackcap's, but without the warbling end-phrase.

Nesting
Garden warblers lay their three to seven eggs in a nest built low in shrub or brambles, tucked away in a shady woodland area.

Length
14cm (5.5in)

Did you know?

■ The garden warbler can be found throughout most of England, but due to lack of suitable woodland it is absent in fen country south of the Wash.

■ Europe, France, Finland and Sweden hold the largest populations of garden warblers.

■ The bird often builds its nest below brambles or nettles for protection.

Because the garden warbler and blackcap songs are so similar, the two birds set up mutually exclusive territories.

The quickest and easiest way to attract birds into your garden is to put out food for them throughout the year. Here, **Peter Jennings from Radnorshire Wildlife Trust** outlines the different ways in which to feed the birds.

Bird tables

Commercially available bird tables can be poorly made and far too small. If you know a handy person with basic carpentry skills, get them to make one for you.

A good bird table should have a feeding area of at least 1m² (2ft²), so that plenty of birds can feed at the same time. Untreated softwood or hardwood, rough sawn or planed, is best. Lengths of 15 x 3cm (6 x 1in) thick board for the top and a length 10cm² (4in²) for the supporting post are ideal. Do not use plywood or any type of composite board. There should be a 1cm-high (0.5in)

ridge around the edge of the tabletop to prevent food blowing away and a 3cm (1in) gap in each corner to facilitate cleaning. This should be done weekly with strong bleach to kill any germs and viruses. Always rinse the tabletop well after bleaching.

The table should be freestanding so that it can be moved about and, ideally, the top should be removable for easy cleaning. Position the table about 1-2m (4-5ft) from the ground and place it in an area where cats and squirrels cannot easily jump onto it. Spread the food all over the table, not just in the middle!

Ground feeding

Many species of bird do not come readily to tables or feeders. Instead, they prefer to feed on the ground, especially on lawns. In fact, all birds will feed from the ground, so if you do not have a bird table or any hanging feeders the birds won't mind! Always scatter food – mixed small seed and porridge oats are best – well out in the open, away from hiding cats, and over as wide an area as you can to allow as many birds as possible to feed at once. This is especially important in cold weather, when an area cleared of snow is very attractive. Try to change the area over which food is scattered often to keep the area hygienic and clean any hard surfaces regularly.

DR S VENITT

Bird feeders

Bird feeders come in all shapes and sizes these days, not just the wire cages and red net bags of the past. They are mostly very good quality and will last for many years.

The mesh cage type is best for dispensing peanuts, and the plastic tube for seed and grain. Some are designed especially for sunflower seeds and others for niger seeds and grain. Most hang from a hook, while others will stick on the outside of a window. Buy the longest ones

you can afford, so that several birds can feed at once, and space them out around the garden rather than all in front of the kitchen window, so that shyer species benefit. You can make your own mesh feeders fairly easily using the smallest gauge of weldmesh.

It's very important to remember that the dregs in the bottom of dispensers should be frequently tipped out and all feeders should be well cleaned and washed regularly to prevent the spread of disease.

Types of food

Nowadays, a huge range of bird foods is available and it has become a very big industry.

■ Most peanuts sold these days are of good quality and are safe for feeding to birds. Avoid buying any that look dry and shrivelled. The smaller varieties are best.

■ Sunflower seeds are mostly black, but the stripy ones are fine. Shelled sunflower seeds (sunflower hearts) are much more expensive, but you do not have to sweep up piles of seed cases and they are easier for birds to eat.

■ There are lots of garden bird seed mixes available. Choose the mix which suits the bird species visiting your garden. Many have a lot of large wholegrains, such as wheat, which few birds eat and which could attract rats if left uneaten, as well as being a waste of money.

■ Fat balls, usually with various seeds in them, are excellent, easy to make and sometimes cheap to buy. Tits love them and they are especially useful in cold weather to get calories into your garden birds quickly.

■ Live and dead mealworms are available, and if you want to get robins onto your hand, live mealworms and patience are the way to do it!

■ Probably the best all-round garden bird food is porridge oats. It is generally cheap, especially if bought in large bags, very few garden birds do not eat it and it is good food for many nestlings.

■ Overripe fruit, apple and pear cores, melon pips and, of course, stale bread and the crumbs from the breadboard and biscuit tin can be fed to birds, but bacon rinds and other meat waste are best put in the rubbish bin, unless you have voracious crows visiting. Feeding the birds can be pricey, but you can reduce costs significantly by buying by the sackful, in partnership with friends or neighbours, perhaps.

Visionary
WETLAND
ideal for birdwatching

Models available in the
Visionary Wetland range:

8x25 **10x25** **8x42** **10x42**

High quality waterproof DCF binoculars. The WETLAND series are designed for birdwatching and nature viewing.
Rubber armoured and fully waterproof, Wetland binoculars are quite at home in the tough outdoors.
High grade roof prisms, coated lenses for great clarity and colour. Rubber eyecups fold for glasses-on use with 8x25 and 10x25, twist on 8x42 and 10x42 models
Includes case, strap and 10 year guarantee.

The full Visionary range comprises of more than 50 different models of binoculars, monoculars and telescopes with retail prices from under £10 up to around £400 plus a huge range of tripods and accessories.

9 new telescopes now available from **Visionary**. Retail prices from around £80 to £400

for more information & to see the full range, visit:
www.visionary.me.uk

When to feed

It is perfectly all right to put out food for birds throughout the year. It is rare that adult small birds try to feed nestlings inappropriate food, such as whole peanuts. So keep your usual peanut and seed dispensers going and put out handfuls of porridge oats all year. It is particularly important not to suddenly stop when the first daffodil appears. Late winter and early spring are times of great shortage of natural food for many birds and a time when the females need all their resources to lay their eggs, and the males to sing and defend a territory. Feeding throughout the spring and summer will generally increase the number of breeding birds in your area and also bring young birds to your bird table and garden, such as spotty young robins, flocks of young tits and, if you're lucky, a family party of great spotted woodpeckers. In the winter it may be necessary to put out food several times a day, but throughout most of the year first thing in the morning is enough, with a lunchtime top-up, if necessary. Try to avoid scattering food on the ground late in the day, as this can attract rats to any leftovers during the night.

■ Just because spring has arrived, don't be tempted to reduce the amount of food you put out for the birds. Natural food resources are often at their lowest at this time of year, while the need for it among birds – now that they have broods to raise – is never greater. Live food in particular can provide the proteins that young birds need in the early days of their life.

■ If you feel uncomfortable about handling live food, such as worms, remind yourself that by providing food for the parents, you'll be helping to ensure that they're fed well enough to search for live food for their young.

■ By providing food during the summer months, you will encourage young fledged birds to stay in and around the general area of your garden. This in turn makes it more likely that they will continue to use it as their base once the colder winter months arrive.

Nick Brown, Derbyshire Wildlife Trust, advises:

■ As a child, I was frightened into believing that chicks would choke to death on oversized pieces of food brought to the nest by overanxious parents. This is quite unlikely. What is more important is the type of food supplied rather than the size of the piece.

■ The traditional feeding of bread has to be done in moderation (as with all diets), and it is critical that when it's offered it is with a plentiful supply of other suitable food. During the peak chick-feeding period (March to July), most young chicks require a balanced diet of high-energy food types usually found in the wild as invertebrates, the favourites being caterpillars or worms.

■ It is no coincidence that blue tits time the hatching of their one and only brood to a bountiful and natural supply of young caterpillars, found high in the unfurling tree canopy. By providing waxworms or mealworms at the bird table during this time, the birds will become familiar with a regular site where food is available, allowing the impact of a wet spring to lessen. Even house

sparrows and chaffinches – traditional seed-eaters – will feed their young on this available source of protein-rich food, which is perfect for the growth of young wing muscles.

■ Between July and March the food selected will vary, but, generally, food with a high fat or oil content will satisfy most birds. Black sunflowers will attract finches while a high-energy seed mix will help tits, sparrows and thrushes through the winter.

■ As at our own tables, food hygiene is critically important. Populations of harmful bacteria

will build up on supplies of food that have not been eaten, leading to (at worst) the poisoning of visitors. It is important to keep the table clean, removing rotting food or piles of uneaten scraps. This can be aided by putting out smaller quantities of food and topping up when supplies have nearly gone.

■ So, without doubt, feeding birds is definitely a good thing to do. Provided with the right type of food at the right time of year, garden birds will thrive on a menu of bird-table delights.

Goldcrest
Regulus regulus

You'll probably hear a goldcrest before you
see it. This tiny flitterer among the twigs and
leaves emits a very high-pitched *seep-seep-seep*
that sounds like a baby bird or even a mouse.
At a distance of more than about 15 metres it
becomes virtually impossible to hear. Once
you've located the small creature, though, you'll

have to keep your eyes fixed, because it twists
and turns through the foliage at a great rate,
even hiding itself behind leaves as it plucks
away the insects that it hunts all day long.

Although most gardens with reasonable
insect activity stand a chance of being visited by
goldcrests, it's the tall stands of coniferous trees

that they particularly prefer. The complicated structures of such trees, as well as the monkey-puzzle, the yew and even the leylandii, give insects plenty of hiding places from most birds, yet the minute size of Britain's smallest resident bird enables it to dig them out.

The goldcrest may be tiny, but it has plenty of spirit. Its most notable feature, the flash of deep orange on its crown, becomes a signal of threat to intruding rivals, the male raising its feathers in a fiery glow, bobbing his wings and calling as loudly as he can. A similar performance is displayed when it becomes time to attract a mate.

Yet perhaps the most amazing aspect of this minuscule bird is its stamina. Late in the year, Britain's resident population of goldcrests is joined by several more that make the long flight, not across the Channel from mainland Europe, but across the North Sea from Scandinavia. Many don't make it, and those that do probably need favourable winds and good weather to help them on their way, but this is such an astonishing feat for such a tiny animal that East Anglians who saw them arrive once thought that they had hitched a ride on the back of an owl!

> 'The goldcrest is not a shy bird but this seems due, not so much to tameness as to indifference: the indifference that small insects show to large things such as human beings.'
>
> From *The Charm of Birds*
> by Viscount Grey of Fallodon

Facts

Frequency
A little under a million pairs breed in Britain each year, and can be found throughout the country although rarely in the northern reaches of Scotland.

Identification
The tiniest resident British bird, complete with its bright pate, can be confused with little other than its close relative, the firecrest. The firecrest is an occasional visitor to Britain, and an even more occasional breeder, and it sports a bold white stripe above its eye which the goldcrest lacks. Young goldcrests look like their parents, but don't have a crest until they've completed their autumn moult.

Song
A high-pitched, thin, pinpoint *seep-seep-seep*.

Nesting
Both parents work together on the nest, a deep basket made of moss, lichens and wool bound together with spiders' silk. Scaffolded as it is with cobwebs, it takes at least two weeks to build.

Length
9cm (3.5in)

The goldcrest weighs about 6 grams – its egg doesn't even weigh one single gram.

Did you know?

■ Goldcrests raise at least two broods each year as insurance against harsh winters which can destroy populations.

■ It's always worth scanning large tit flocks for the possibility of a goldcrest that's come along for the ride.

Goldfinch
Carduelis carduelis

Usually, flowers provide the colour in a garden, while birds supply the entertainment. When goldfinches drop by, you get the best of both. At first, all you may hear is a faint cheery tinkling sound, rather like far-off wind chimes, as they call to each other while flitting from one food source to another. Then the flock bounces into view in a blur of crimson and gold. As they start to settle on a bird feeder or seedheads, you get a chance to admire their beautiful gaudy plumage: the bright red, white and black headdress, fawn-brown body and golden trim on jet black wings.

For much of the year, flocks of goldfinches roam the countryside in search of seeds. In early summer, a goldfinch usually targets the fluffy seeds which have parachuted away from ripe thistleheads and landed on the ground – goldfinches seem to prefer their seeds slightly underripe. Any seeds left on the plant are plucked later in the year as the bird clings on to swaying seedheads and bendy stems. Unlike most songbirds, the dainty goldfinch can use its feet to pull a seedhead nearer and hold it steady while it extracts the seeds. It often lands low down on a stem, then moves up so the seedhead bends over and comes within reach of a grasping foot.

In autumn, many British goldfinches, mainly females and juveniles, shoot off to France and Spain. At the same time, refugees from chilly northern Europe start arriving along the east coast to spend the winter here. Leaving part of your garden to run wild with thistles, groundsel,

> 'Sometimes goldfinches one by one will drop
> From low-hung branches; little space they stop;
> But sip, and twitter, and their feathers sleek;
> Then off at once, as in a wanton freak;
> Or perhaps to show their black and golden wings
> Pausing upon their yellow flutterings.'
>
> From 'I Stood Tip-toe Upon a Little Hill' by John Keats

coltsfoot, ragweed, dandelion, knapweed and burdock would give you a colourful patch during the summer and encourage goldfinches to pop by for the seeds, especially in the spring when the migrants return to find wild seed stocks running low. They also garden-hop to pinch husk-free sunflower seeds, millet and tiny black niger seeds from hanging feeders or low wire trays.

Did you know?

■ A party of goldfinches is known as a 'charm'. While they are very charming, the name was actually derived from the old French word for song, *charme*, as a tribute to their bell-like tinkling.

■ Goldfinches are the only finches with beaks long and fine enough to extract teasel seeds from deep pockets in the spiky seedheads, although it's mostly males that feed on them as they have slightly longer beaks than females.

■ The goldfinch's exotic plumage and sweet singing used to cost millions their freedom as the highly prized victims of a flourishing trade in caged birds. In the autumn of 1860, local records reveal that as many as 132,000 goldfinches were trapped around Worthing on the south coast of England as they made their annual pilgrimage to the Continent for the winter. In spite of the Protection of Birds Act (1880), the illegal trade in wild songbirds went on until the Wild Bird Protection Act in 1934 largely eradicated it (or forced it undercover, as there's still a small black market in goldfinches).

Facts

Frequency
Most visible in gardens during the spring when they are patrons of seed hangers and trays; population of around 230,000 breeding pairs drops to about 100,000 over the winter. Numbers have partially recovered from a sharp drop between the mid-1970s and mid-1980s, probably caused by herbicides, which cut the supply of weed seeds on farmland, or excessive shooting and netting on the Continent during the winter. Being fed in gardens has probably contributed to their recovery.

Identification
Unmistakable fawn plumage with red, black and white headdress and bright gold bands on black wings; the red on the male's head extends further behind the eye; juvenile is streaky dull grey-brown, lacking any colourful markings on its head until after a partial moult in autumn.

Song
Sweet tinkling melody.

Nesting
Female weaves fine roots, grasses, moss and lichens into a deep, thick-walled cup, anchored to the twigs on the upper branches of a tree.

Length
12-13.5cm (5in)

When probing for teasel seeds, a goldfinch makes a curious buzzing *geez* sound. As the bird pokes its beak down into each pocket, such vibrations may help to dislodge the seeds, making them easier to pinch in its beak and withdraw.

Bathmate's MAGIC Air-cushion!

"It's so easy!"

Bathmate is your safest, simplest way to enjoy a proper bath - anytime, anywhere. Bathmate's **UNIQUE** air-cushion forms a comfortable seat and back rest.

IN
Deflates... easing you gently down...

RELAX
Then lie back safely and luxuriate...

OUT
At the touch of a button...up... safely out...

BUY WITH CONFIDENCE FROM A COMPANY THAT CARES!

- Fully portable with its own handy travel case
- Slip resistant seat pad • UNIQUE air filled back support cushion • To ensure your complete satisfaction we offer a FREE home trial • Suits most baths - even small ones • No installation or fitting required

For a FREE brochure **FREEPHONE** **0800 072 9898** ASK FOR EXT. 70839

Or write to: Freepost Nationwide Mobility.   Or visit: **www.nationwide-mobility.co.uk**

Please send me a FREE colour BATHMATE brochure   70839

Name _____ Tel No. _____

Address _____

_____

_____ Postcode _____

Send the coupon to: Freepost Nationwide Mobility.

Nationwide Mobility

Stephen Hussey, Devon Wildlife Trust, advises:

■ A succession of thin tinkling calls are often the first sign of goldfinches. Look up and you will see parties, or charms, passing in a bouncing flight pattern of rapid wing beats, followed by short glides with wings closed.

■ The same features that make goldfinches such a welcome sight in our gardens today were once the birds' undoing. Exotic-looking plumage combined with a pleasing song to make it the ideal caged bird for Victorian collectors. At a time when no home was complete without a tamed songbird, hundreds of thousands were being caught and traded each year.

■ Today the goldfinch is faring better. The reinstatement of field margins and other 'waste' spaces in parts of the farmed landscape is providing a supply of their favourite food: seeds. The bird's scientific name *Carduelis carduelis* means 'of the thistle' and is apt. The goldfinch's small size allows it to balance atop agricultural weeds, such as teasels, thistles and knapweed, probing them with, what is for a finch, a slender beak.

■ Goldfinches have become a more common sight in rural and urban gardens over the last 20 years. However, they still possess a residual nervousness around people and are often the first birds to leave a garden when its owners appear.

■ Including any of the goldfinch's favourite seed plants in your wildlife garden should tempt them to visit. But if growing weeds seems a step too far in luring these charismatic birds, garden favourites such as aster and pansy will substitute. Goldfinches rarely visit bird tables, but parties will happily queue for a spot at a hanging feeder filled with their favourite niger seeds.

Great-spotted Woodpecker
Dendrocopos major

Did you know?

■ Woodpeckers are the only birds in Britain to use instrumental sounds as well as vocals in their courtship routine each spring. Amorous male great-spotted woodpeckers in particular are keen on beating tattoos on hollow branches with their beaks to send out low-pitched invitations, which carry a long way through the trees, to attract a mate. Each bird plays his own unique signature tune as he vies to be top of the would-be pops.

■ There is another, less common black and white woodpecker, the lesser spotted woodpecker (*Dendrocopos minor*), living in British woodland and parks. Although the two species of woodpecker look superficially similar, the great-spotted is bigger, with prominent white shoulder patches and a striking red patch under its tail, which the lesser spotted lacks. The male great-spotted also has a scarlet patch on the nape of its neck, whereas the male lesser spotted has a scarlet crown.

■ Another way of telling the two apart is to watch the back in flight – the great-spotted has a plain black back, while the lesser has white bars across its back.

'He walks still upright from the Root Meas'ring the Timber with his Foot; And all the way, to keep it clean Doth from the Bark the Wood-moths glean.'

From *Upon Appleton House* by Andrew Marvell

If you live in a tree-rich setting, you may get regular visits from the black and white bird with a scarlet flash under its tail. But for other country dwellers, a glimpse of a great-spotted woodpecker hopping up a tree trunk or bashing away at a peanut feeder is pure serendipity.

You soon get an idea of how well adapted it is to its tree-based life as you watch it skilfully using its powerful beak like a crowbar to cast aside patches of lichen and moss or lever up loose flakes of bark to expose any insects or spiders living underneath. On dead or dying trees, you may see it pause to listen for movement under the bark, or poke its extraordinarily long tongue into a tunnel made by a wood-boring insect. The tip of the tongue is sensitive, bristly and sticky and can be extended at least 4cm (1.5in) beyond its beak to search for any grub in residence before dragging it from the hole and into its mouth.

To help it climb trees and hold itself steady when drilling, the great-spotted woodpecker's tail feathers have especially thickened shafts. These make the tail stiff enough to form a strong brace to prop up the bird as it leans back to get more power into its pecking or better purchase on bark. Its short legs hold it close to the trunk and, unlike most small birds, woodpeckers have a two-forward, two-back arrangement of toes, each tipped with sharp claws, to clamp it to the bark.

When pecking out a nesting hole in the trunk of a dead or rotten tree, the woodpecker bashes into the timber with its beak, hammering the trunk 15 or 16 times in one-second bursts, rather like a road-digging pneumatic drill. To avoid serious brain damage, there is a shock-absorbing cushion between beak and thickened skull to lessen the potentially damaging effects of constant jarring. The woodpecker also has extremely strong neck muscles, which hold its head steady and aligned, to prevent any slewing that could bruise the brain.

Facts

Frequency
The most common woodpecker living in Britain (there are no woodpeckers on the relatively treeless island of Ireland); doing well, largely thanks to a run of mild winters and the growth in garden peanut feeders. Landowners and woodland managers are encouraged to leave some dead and rotting trees standing in woodland to support its tree-oriented lifestyle.

Identification
Bold black and white markings with a scarlet patch under the tail; the male has a scarlet flash on the back of his neck, but the female's nape is plain black; juveniles have a scarlet skullcap.

Song
A strident *tchack, tchack* call echoing through the woods gives it away before you ever catch sight of it.

Nesting
A pair takes two to three weeks to hack out a roomy nesting chamber in a hollow tree trunk.

Length
22-23cm (9in)

Just add birds

*Welcome feathered friends, old and new to your garden
with our range of quality bird care products.*

David North, Norfolk Wildlife Trust, advises:

■ If you are lucky enough to have a 'great spot' visit your bird table then you will realise what a striking and beautiful bird it is. Its smart black and white plumage, the bright crimson patch under the tail and the characteristic alert upright woodpecker posture give it the air of a visiting celebrity. Other visitors to your bird table seem to show a certain respect when a woodpecker pops by, perhaps in deference to that powerful woodpecker beak.

■ Great-spotted woodpeckers are noisy birds. The territorial drumming of the male is an early sign of spring, often heard as early as February. There is no other British bird that can beat its head so hard against a tree and repeat it several times a second!

■ The good news is that gardens are increasingly being visited by great-spotted woodpeckers attracted by fast food in the form of fat balls and peanuts.

Large gardens with mature trees can also provide breeding sites. Great-spotted woodpeckers excavate their nest holes in the trunks or larger branches of mature trees, usually several metres above the ground. However, even small suburban gardens can become regular stop-offs for great-spotted woodpeckers nesting nearby, as long as suitable food is regularly supplied in feeders. Gardeners can also encourage woodpeckers by ensuring some dead wood is left standing where this is safe to do.

> **Hardly surprisingly, a great-spotted woodpecker generates masses of sawdust while drilling a nesting hole in dead timber. To prevent its airways getting clogged, its nostrils are narrow slits, guarded by wiry feathers to trap the dust. Most of the sawdust gets chucked out of the hole and collects on the ground under the tree, but a few chippings are left on the floor of the nesting chamber to soak up the chicks' droppings.**

Great Tit
Parus major

You see and hear a lot from the great tits in your garden. They're bold in colour, markings and spirit; they broadcast running commentaries on their moods and intentions, aggressive and amorous; and they are forever showing off their acrobatic skills on the peanut feeder.

In a bird-eat-bird world, the inquisitive great tit is better equipped than most of its rivals to finish near the front in the race for food and nesting spaces. If you watch a great tit closely as it goes about its business in your garden – and, fortunately, its yellow breast and black and white head make it easy to follow – you start to appreciate just how smart and ingenious it

is: as it scours a tree for insects to eat, it checks the underside of the leaves for shadows of caterpillars, which is a clever way of spotting them as they are normally green and well camouflaged against the foliage. To save time, it taps an acorn with its beak to assess whether it sounds hollow and is worth opening for the grub inside.

The great tit's never-ending quest for food reaches fever pitch when chicks come along in May and June. (The arrival of a family should be timed to coincide with the hatching of millions of pesky leaf-nibbling caterpillars – parent great tits couldn't keep up with the insatiable

The great tit's bill is thicker and stronger in winter, when it is eating plenty of tough-cased seeds. In the spring, a great tit does lots of bill-wiping, stropping its bill from base to tip along a perch, until it is finer and in better shape for feeding on insects.

Facts

Frequency
One of the most widespread, common and familiar garden birds; over two million pairs breeding every year.

Identification
Male and female have a black head with white cheeks, yellow underparts with a black stripe down the centre and green upperparts with blue-grey wing and tail feathers; males are larger than females, with glossier, brighter plumage and a wider, longer belly stripe; juveniles are duller, greyer and yellower.

Song
Never silent for long; chatters all the time with regular, rapid bursts of reedy, high-pitched *tsee-tsee* contact notes. The male's spring song, *tea-cher, tea-cher*, sounds rather like a wheezy bicycle pump.

Nesting
Generally nest in tree holes or nesting boxes; the female collects strands of dry grass and moss to line the hole, and shapes a small cup which she lines with hair and feathers.

Length
14cm (6in)

appetites of their large families unless there was an ample supply of protein-rich food nearby – although in recent years there have been signs that climate change is pushing the two breeding cycles out of sync.) Stand anywhere near a nest box occupied by a family of young great tits and you soon learn that parent great tits assess the hunger of their chicks by how noisily they demand to be fed, and fetch enough provisions to silence the rumpus.

When the weather becomes cooler, the days shorter and insects scarcer great tits start eating more seeds and nuts. They also visit bird tables, feeding stations and hanging feeders more often, where they have a reputation for bullying blue tits and other small birds out of their fair share of the rations. A great tit simply barges in and makes off with the best titbits and will even raid the carefully laid-down food stores of coal and marsh tits.

'If you would happy company win
Dangle a palm-nut from a tree,
Idly in green to sway and spin,
Its snow-pulped kernel for bait; and see
A nimble titmouse enter in.'

From 'Titmouse' by Walter de la Mare

Did you know?

■ One way of telling a male from a female great tit is to look at the bird's black belly stripe: the male's is wider and extends right down between his legs. In fact, as far as the females are concerned, the wider his belly stripe the better; males with the wider black stripes appear to make better providers for their families. The female has a fine, broken grey belly stripe, which stops short of her legs.

■ The great tit is the largest European tit.

Stephen Hussey, Devon Wildlife Trust, advises:

■ Great tits are a feature of most UK gardens. Their ability to adapt to all but the highest and northernmost parts makes them the most widely recognised of the tit family alongside their smaller cousin, the blue tit.

■ The great tit population has grown steadily over the last 50 years to a total of 1.6 million pairs. The good work of human beings has contributed to this.

■ Great tits have prospered from people's increasing appreciation of garden wildlife. Although primarily an insect eater, great tits have readily taken advantage of the growth in garden bird feeding. They will happily visit bird tables to supplement their wild diet with fruit, seeds and scraps – sunflower kernels are a favourite. Great tits also show enthusiasm for hanging peanut feeders, although they fail to match the blue tit's acrobatic panache.

■ The popularity for providing garden nest boxes has also aided the great tit. Where once small holes in trees served a purpose, great tits have discovered an increasing supply of 'tit' boxes, ie those with a small hole around 28mm in diameter. If positioned well – at or above adult head height and amongst cover – the box should play host in April and May to a great tit brood of up to 12 young.

■ Their numbers and relative confidence around people make great tits fascinating garden birds to watch. Look out for their distinct bird table etiquette where dominant individuals, usually males, will chase away subordinates and other species. Another familiar sight comes during courtship where males will ply their partners with food. With every gift the female will crouch with its gaping mouth raised and wings fluttering, aping the poses that her fledglings will take later that same spring.

Greenfinch
Carduelis chloris

If you could transport yourself and your garden back to the 19th century, you'd notice that something familiar was missing. A century or more ago, greenfinches were extremely rare garden visitors, restricting themselves to the countryside only. As the suburbs grew, however, unlike most species which retreated further into rural areas, greenfinches adapted very quickly, and soon came to rely more and more on the food put on the bird tables of the nation. Today, many greenfinches use gardens for the majority

Did you know?

■ Greenfinches adapt quickly. They were first recorded eating the fruit of *Daphne* garden shrubs in the 1930s, and the practice became commonplace across the country within just 30 years.

■ The bird is widespread across Europe, and can even be found as far afield as China and Japan.

■ Greenfinches have been recorded as living for up to 13 years.

Young greenfinches are like female house sparrows, so look for the yellow flashes on wing and tail to be sure.

Frequency
The 700,000 breeding pairs that live across all of Britain (bar the northern parts of Scotland) are augmented in the winter by at least two million visitors from the continent.

Identification
In full summer plumage, the male greenfinch displays an olive-green back and lighter green underparts, while the female is generally duller with streakier markings. After the autumn moult, the birds become much browner during the winter, but the yellow splashes on their wings and the sides of their tails help to identify them. This bird also has the stockiest bill of the common garden finches.

Song
During the breeding season, males sing a drawly, wheezing song, sounding as if they have breathing difficulties. To keep in contact with each other, flocks twitter among themselves.

Nesting
Hedges and thickets are the favoured breeding spots of the greenfinch, which builds an untidy nest of grass, wool and moss, occasionally peppered with the odd feather.

Length
15cm (6in)

of their food, particularly during the winter months, and over 1,000 different individuals have been recorded passing through a single garden during this season. During the harsh winter of 1962/3, peanuts from bird tables made up to 97% of their food.

These adaptable birds have made gardens their own, and they like to make sure they keep it that way. They are very jealous of other birds that approach while they're at the table or the feeder, and their stockiness gives them the edge over tits and other finches.

Yet greenfinches aren't completely reliant on artificially provided meals. Mezereon and laurel berries are a particular favourite, and if you have sunflowers, it's worth postponing deadheading them until the greenfinches have completely cleaned out the seeds. Because of their size and bulk, however, they tend to go mostly for seeds that have fallen to the ground, making them a great help as weed seed clearers for gardeners.

Greenfinches do have one clever trick up their sleeve, though. Unable to prise the seeds from closed pine cones, they wait until the cones open up on a sunny day, before plucking them out.

'Its note sounds like "eeze" or "wheeze"; and I think of it as a welcome suggestion on sultry hot days.'
From *The Charm of Birds* by Viscount Grey of Fallodon

Green Woodpecker
Picus viridis

An olive-green swoop across your garden, a flash of red, and a maniacal laugh – it's enough to brighten any day. Green woodpeckers are among the most impressive of British birds, their sturdy frame rising and falling as they undulate through the air. Alternately flapping to gain height, then coasting with closed wings, they actually appear to be bouncing across the sky.

And then there's that laugh. The ringing tones of the woodpecker's call have given it several nicknames in various parts of the country over the years – heigh-wawe, highhoe and yaffle being just a few. A shy bird, that tends to scamper around the back of a tree to avoid being detected, its presence tends to be given away by that echoing cackle.

Unlike the great spotted woodpecker, the green woodpecker is unlikely to visit your bird table (other than in very tough winters when it may pick at suet cakes), but this doesn't mean it will keep away from your garden. If you live near trees, and have a reasonably sized lawn, you have a chance of seeing one drop in, settling on the grass, and busying itself in a hunt for ants.

The green woodpecker's tongue is very long, reaching about 10cm (4in) beyond the tip of its bill. It uses it to stab at ants that are trying to tuck themselves into the earth. A few jabs of its bill into the ground, a cocked glance at what's revealed, and then out flicks that extraordinary tongue and the ant simply disappears.

Greenies don't just come to gardens for the insect life, though. If you leave apples where they've fallen, and perhaps provide a few nuts into the bargain, the bird may well take you up on the offer, particularly during colder winters when invertebrate life is harder to come by.

> **'Under the crag, where the tree-tops lean**
> **Flashed your feathers in green and gold'**
> From 'The Yaffle' by AC Benson

Facts

Frequency
The green woodpecker can be found across most of Britain, except northern parts of Scotland, and numbers some 15,000 breeding pairs. The bird is absent from Ireland.

Identification
A scarlet crown, dark green back and black face mask, streaked with a red moustache in the case of the male, are the main points of identification. In flight, look out, too, for the bright yellow rump. The long dagger-like bill should help clinch the tick.

Song
An unmistakable *hew-hew-hew* laugh, which descends the scale and rings through the trees. Similar notes are used in the flight call.

Nesting
Although the green woodpecker can excavate a hole for itself in rotting wood, it prefers to take over existing holes that have perhaps been used by other birds in previous years. Rather surprisingly, given its size, the green woodpecker is sometimes ousted from its nest site by starling pairs, which bully the giant until it leaves.

Length
32cm (13in)

Ants aren't the only social insect on the woodpecker's menu. In cold winters, they've been known to dig out bee and wasp nests, and even break into bee hives. Their apparent invulnerability to stings led to the old belief that beekeepers would be protected if they carried a woodpecker's beak.

Did you know?

■ Many sightings of the rare golden oriole in Britain are actually a case of mistaken identity, the yellow flash of the green woodpecker's rump leading to the confusion.

■ A crafty woodpecker will hunt ant trails on the woodland floor, then simply scoop them up as they march towards them.

Insects are an essential part of the food chain in gardens. Many birds' diets consist solely, or partly, of insects, and they are a particular favourite when adult birds are feeding their young. However, insects often get a bad press. Some sting or suck blood, while others damage plants by sucking sap or eating the foliage, roots, flowers or fruits, but only about 1% of the 22,000-plus insects in Britain are significant garden pests. **Andrew Halstead of the RHS** shows that the vast majority of our insects do no harm and some are of real benefit in gardens.

Many insects visit flowers to feed on pollen and/or nectar. In doing so, they frequently pick up pollen on their bodies. When they visit another flower of the same sort, that pollen may get transferred to the stigma of the flower, allowing pollination to occur. Bees are particularly good as pollinators, as their hairy bodies readily pick up pollen grains. Most fruits grown in gardens, as well as many vegetables and ornamental plants, are dependent on insects for pollination. Honeybees, of course, also provide us with honey and beeswax.

Some insects feed on other insects and invertebrate animals, several of which are plant pests. Examples of insects that prey on aphids include ladybirds, some hoverfly larvae and lacewing larvae, while ground beetles, or carabids, and rove beetles, or staphylinids, feed on a wide range of soil-dwelling pests.

Insects such as ichneumon, chalcid and braconid wasps, and tachinid flies, lay their eggs in the eggs, larvae or pupae of various insects, especially moth and butterfly caterpillars. The parasitoid larvae develop inside the bodies of their host insects, often feeding for several weeks before the host insect is killed. One wasp, *Encarsia formosa*, is used as a biological control of glasshouse whitefly, while *Aphidius colemani* is used against aphids in glasshouses.

Insects are also a significant part of the diet of many other creatures. Spiders, amphibians, birds, bats and other mammals rely on insects for all or most of their food requirements. The pivotal role of insects in many food chains makes them a vital part of garden biodiversity.

Insects, especially some flies and beetles, feed on dead plant material, helping to break it down and making it available to growing plants as humus and nutrients. Insects also play an important role in disposing of animal dung and the corpses of birds and mammals.

113

Making your garden insect-friendly

Gardens can be regarded as a collection of habitat types, each of which will appeal to certain kinds of insects. To get you going, here are some of the more important habitats for insects.

Flowers

Flowers provide a service to insects that is similar to that of petrol stations to cars. They provide a high-energy fuel in the form of nectar, with the added bonus of protein- and oil-rich pollen grains. Day and night, flowers will be visited by a steady stream of butterflies, moths, flies, beetles, bees and other insects. Plant your garden so that there is something in flower from early spring to late autumn. Mixing flowers with vegetables and fruit trees may encourage hoverflies to visit and lay their eggs on the fruits and vegetables where aphids can be a problem.

Wildflowers

Some native wildflowers are sufficiently attractive to earn a place in gardens (see page 54 and onwards). Exotic plants can attract a wide range of insects but these are mostly non-specialist insects that can use a wide range of plants. By growing native plants in a garden you can encourage those specialist insects that are entirely dependent on those plants. It is best to select wildflowers that already occur in your area. This links your garden to the wildflowers and insects of the locality.

Ponds

Adding water to a garden introduces a range of insects that cannot breed in a 'dry' garden. Dragonflies, damselflies, pond skaters, water boatmen, caddis flies, mayflies, water beetles, flies and other insects are dependent on water and pond-side vegetation. Omitting fish from the pond will increase the diversity and abundance of insects. Even small bodies of water, such as a washing-up bowl with a thin layer of soil under the water, will allow an insect community to develop.

Lawns

Fine, frequently-mown lawns offer little to insects. However, lawns where clovers, daisies, dandelions, bird's-foot trefoil and other short-turf wildflowers are allowed to grow and flower can be attractive to both the eye and insects. Where space allows, part of the lawn can be allowed to grow taller to accommodate meadow flowers. Long grass will suit grasshoppers, crickets, plant bugs, skippers and other meadow butterflies.

Compost heaps

Gardens generate leaves, grass mowings, prunings, dead flower stems, leaf mould and weeds that need disposal. Taking them to a tip is a loss of valuable organic matter from your garden. They should be rotted down in a compost heap or bin to be used as a mulch, or dug into the soil. Compost heaps teem with the larvae and adults of beetles, flies, springtails and other invertebrate animals, which contribute to the composting process by breaking down the plant tissues. In the autumn, resist cutting down and composting all of the dead stems and foliage on herbaceous plants. Leave some until late winter or early spring to avoid flat, empty flower beds and to provide over-wintering sites for insects.

Hedges

Fences and walls support few insects, so planting hedges should be considered where this is feasible. Broad-leaved hedging plants support more types of insect than conifer hedges, although the latter make good winter shelters. Hawthorn, privet, hornbeam, beech and holly are suitable for clipped hedges, but regular clipping will restrict the plants' ability to produce flowers and fruits. An informal hedge or screen, where the plants are infrequently cut back, is better

for most wildlife, including insects. Planting with a mixture of hedging plants, instead of a single species, enables the hedge to support a wider range of insects. An informal hedge can include some scrambling hedgerow plants, such as wild rose, honeysuckle, white bryony and bramble. If you haven't got space for a hedge, plant climbers, such as ivy and honeysuckle, against your wall or fence.

Dead wood and rot holes

Safety considerations may require the removal of dead trees or branches, but where possible, consider leaving dead wood in place. Dead wood is far from lifeless in terms of value to insects. About 20% of woodland insects are directly or indirectly dependent on dead or dying trees. A wide range of beetles and flies feed on decaying wood or associated fungi. Some solitary bees and wasps reuse beetle tunnels as nest sites, or make their own in rotten wood. When trees need to be felled or branches removed, the larger timber can be stacked, preferably in a shaded area, for the benefit of dead-wood insects and other wildlife. Standing trees sometimes develop rot holes.

Storm damage or bad pruning exposes the inner wood and allows colonisation by fungal decay that gradually results in a cavity forming in the trunk. These fill with rainwater, dead leaves and the corpses of birds and small mammals. The nutrient-rich sludge at the bottom of rot holes is home for some of Britain's rarest beetles and flies.

Bee nests

Many solitary bees and wasps nest in beetle tunnels in dead or rotten wood, or in hollow plant stems. Artificial nest sites can be created by drilling holes of a variety of diameters, 2-9mm, in fence posts or blocks of wood. Hollow stems of herbaceous plants or bamboo canes can be pushed into plastic drink bottles from which the tops and bottoms have been removed. Place the nests in sunny places under the eaves of a shed or on the rail of a fence in the spring. Boxes of nest tubes and drilled blocks of wood are often sold in garden centres. They are also available by mail order from:
CJ Wild Bird Food,
(www.birdfood.co.uk)
or Agralan Ltd
The Old Brickyard, Ashton Keynes, Swindon,
Wiltshire SN6 6QR,
tel: 01285 8660015,
(www.agralan.co.uk).

Pesticides

Insecticides used to control plant pests will also kill the insects that you are trying to attract. Insecticides should only be used as a last resort. Avoid spraying all the plants 'just in case' they might have pests. Organically approved insecticides based on natural substances, such as pyrethrum, rotenone/derris, fatty acids or plant oils, have little persistence, but even they will harm beneficial insects. If you do use an insecticide, make sure you spray at dusk when bees are not flying. Biological controls, using predators or parasitoids, are available for many glasshouse pests, and for leatherjackets and chafer grubs in lawns. If used correctly, biological controls can be effective alternatives to pesticides. Further advice on encouraging garden biodiversity is available on the RHS website, www.rhs.org.uk/biodiversity.

Agralan offer a wide range of non-chemical plant protection products, including biological controls.

Tel 01285 860015
to request a catalogue
or visit our website.

www.agralan.co.uk

Create a butterfly haven

The Berkshire, Buckinghamshire and Oxfordshire Wildlife Trust advises:

■ Butterflies are well known as attractive creatures, but many other insects are worth a second glance. A close inspection of insects on plants, and watching their behaviour, can greatly add to the interest of a garden.

■ Butterflies such as red admirals and small tortoiseshells are attracted to rotting fruit. Place a few bruised apples around your garden, as an addition to nectar plants.

■ Avoid using insecticides in your garden as these may reduce butterfly numbers. Try to use alternative forms of control by encouraging natural predators, such as frogs and hedgehogs.

■ Provide sheltered, undisturbed areas in your garden where adult butterflies and caterpillars can hibernate over the winter months. These can include ivy and piles of logs and leaves.

■ Make a butterfly feeder! Stick a milk-bottle top to a piece of brightly coloured card. Fill the bottle top with a sugar solution made from one teaspoon of sugar to 20 teaspoons of water. Place the feeder in a sunny, sheltered spot in your garden and watch the butterflies feed. Alternatively, you could soak a piece of material in the sugar solution and hang it from a tree or fence.

Plants for insects

Private gardens in Britain cover about 270,000ha (667,185 acres), writes **Andrew Halstead of the RHS,** and together with public parks, provide important feeding and nesting areas for bees and butterflies – the pollinators of gardens.

Habitat features, such as hedgerows, ponds, lawns and flower borders, can be put into a garden to encourage insects. Some native wild plants are attractive enough to earn their own place in gardens and may help support bees and other insects that otherwise could not survive in urban areas. It should be remembered that bees like the sunshine and may ignore wild plants hidden in shady corners. Care should be taken to ensure that purchased wildflower seeds or plants are of native stock and not strains of the same plants obtained from other countries, as these may have different flowering periods and growth habits compared to native plants.

Many garden plants are of exotic origin or have been hybridised to produce flowers with different characteristics to the true species. These can be of value to bees with non-specialised feeding requirements, but may be unattractive to wild bees which are restricted in their flower choice. These include monolectic species (which visit just one plant) such as *Andrena florea* and *Macropis europaea*, which only take pollen from white bryony (*Bryonia dioica*) and yellow loosestrife (*Lysimachia vulgaris*) respectively.

There are also other bees (oligolectic species) which confine their pollen-collecting activities to a few closely related plants, for example *Andrena apicata* on willows (*Salix* species) and *Andrena lathyri* on vetches (*Vicia* species). Polylectic bees collect pollen from a wide range of flowers; examples of these are the honeybee and bumblebees.

Various annual and perennial wildflowers can be incorporated into borders and will not look out of place among the more conventional garden flowers. They will often make bigger plants than in their natural surroundings, where competition

with grass and other wildflowers restricts their growth. Adult butterflies visit flowers to obtain nectar, which they suck up with their long tongue, or proboscis. They make use of a range of flowers and other sweet substances, such as the juices of overripe fruits. The plants listed (right) will attract butterflies, but to get the best results they should be grown in sunny, sheltered places. Clumps of plants are usually more attractive than single scattered plants. Nectar-providing flowers are also used by other insects, such as bees and moths.

Insect-friendly gardening

Gardens can be made insect-friendly in various ways:

■ Insects visit flowers day and night to feed on energy-rich nectar and pollen. Plant your garden so there are plants in flower from early spring to autumn. Include some of the more attractive local wildflowers, as this provides a link to the wider countryside.

■ Adding a pond, or even a water-filled washing-up bowl, will attract water beetles, pond skaters, dragonflies and other pond life.

■ Frequently-mown fine lawns support few insects, but wildflower lawns where clovers, daisies, dandelions and bird's-foot trefoil are allowed to flower are much better. If grass is allowed to grow tall, meadow wildflowers can be grown. This will suit grasshoppers, crickets, skippers and other meadow butterflies.

■ Gardens need a compost heap for recycling waste plant material. Compost heaps teem with insect life, especially springtails, beetle and fly larvae that help the composting process.

■ Hedges are better for insects than walls or fences. Informal hedges of mixed plants, including scrambling plants, such as wild rose, bramble, white bryony and honeysuckle, are better than a single-species hedge.

■ Dead wood is far from lifeless in terms of value for insects. About 20% of woodland insects directly or indirectly live on or feed on dead wood and the associated fungi. When trees are felled or pruned, use the trunks and larger branches to make a log pile in a shady corner of the garden.

■ Provide nesting sites for solitary bees by drilling holes of diameter 2-9mm in fence posts or blocks of wood. Bundles of hollow herbaceous plant stems or bamboo canes can also be used.

■ Keep pesticide use to a minimum. Biological controls, using natural enemies, are available for some pests. Short-persistence pesticides, containing pyrethrum, derris/rotenone, plant oils or fatty acids, are less likely to harm non-pest insects.

Grey Heron
Ardea cinerea

Herons in the garden create something of a dichotomy. If you have fish in your pond that you're particularly proud of you won't want herons, but they'll come in anyway. If you don't have any fish and would love to see a heron, well, you'll probably be disappointed.

Herons like fish, and with the gardens of England providing plenty of opportunities for finding them, it's not surprising that the biggest bird you're ever likely to see dropping in for a visit is this wary, patient giant that thinks nothing of waiting for hours alongside a body of water, waiting for the right moment to strike. It is sometimes easy to miss the ghostly figure, but once disturbed, its large wings, deep wingbeats and distinctive shape with head tucked back onto the neck are unmistakable.

When hunting, grey herons appear to be solitary birds, but they nest communally in tall trees, and form quite a society. The nests are refurbished and used every year with the experienced males claiming the best sites, while new or younger males usually have to make do with what's left or build anew on the outskirts of the colony. Colonies also have 'club' areas in nearby trees nearby where young, off duty, immature birds can practise their courtship displays.

Herons swallow their prey whole but, in the breeding season, they will carry the meal back to the heronry in roomy pouches in their neck.

> **'The moping heron, motionless and stiff,**
> **That on a stone, as silently and stilly,**
> **Stood, an apparent sentinel, as if**
> **To guard the water-lily.'**
>
> From 'The Haunted House' by Thomas Hood

Herons live in the UK all year round, although unusually harsh winters or dry summers have been known to force them across the channel.

Facts

Frequency
Some 13,000 pairs breed the length and breadth of the British Isles.

Identification
Male and female look alike, upperparts grey, underparts whiteish, while the long neck is white with black streaks. The white head has a wispy black crest. The dagger-like beak is yellow but turns light orangey-pink during the breeding season. The juvenile is darker and greyer with no crest.

Song
A harsh *fraank, fraank* call, and a rapid bill-clacking at the nest.

Nesting
Twiggy baskets in tall trees, often in groups of 30-50, although up to 100 have been recorded in one heronry.

Length
90cm (35.5in)

Very young chicks get regurgitated fish soup down their neck, but once they are older the food gets dumped in the nest while the parent flies off for another round of hunting. Parents take turns hunting when they have eggs or very young chicks, and the change of duty entails a complicated greeting ceremony. There's much puffing out of breast feathers and pointing of necks and heads skywards, then some bill-sparring completes the ritual, and it's off to find some food again.

Whether you love or loathe herons, they are an encouraging sight. As the birds like to hunt in clear water, their presence is a good indicator of the health of a body of water and its ability to support plenty of fish life. A good sign that all is well with that particular environment.

Did you know?

■ In falconry the grey heron was considered a royal bird, and only kings could hunt them. The usual practice was send a number of falcons after the big birds in the hope that the raptors would stoop down and stun them. Herons became very efficient at avoiding death, though. They would fly as high as they could thus not allowing their assailants enough scope for a stoop, and if this didn't work, they would vomit up their latest meal to lighten themselves and make a quick getaway.

■ The heron is the tallest resident British bird. It's quite long-lived, too, capable of living for up to 25 years.

Hobby
Falco subbuteo

This is a fascinating story, which just happens to be true. Several decades ago, a games enthusiast invented a tabletop entertainment that he felt sure would catch on. He took his idea to the patent office and asked to copyright it under the name 'hobby'. 'Sorry,' he was told, 'hobby is too generic a name to be able to be copyrighted.' Somewhat disappointed, he discussed this with a friend, who pointed out that hobby was also the name of a bird of prey. 'Why not find out what its scientific name is, and call your game after that?'

And that is how Subbuteo, the long-popular tabletop football game, came into being. The word itself actually means 'sub-buteo', or 'small buzzard', which is a bit of misnomer because

this fascinating summer visitor is actually a falcon. And what a falcon it is. Its sharply pointed wings enable it to twist and turn with the best of them, and it's actually able to catch that extraordinary aerial acrobat, the swift. Swallows and martins, which are slightly less agile, form a larger part of the hobby's diet, though.

So why is this bird included in a book about gardens? The answer is that, although it is highly unlikely to settle in your garden, if you have a sizeable pond with plenty of dragonfly activity, on a bright summer's day, when the insects are flying free and high, you might be treated to an amazing display in the skies. The sight of a hobby, with its russet-brown trouser-like leg feathers swooping round and round and plucking zigzagging dragonflies from mid-air is one of the great sights of nature. The chase is often quite quick, otherwise the dragonfly tends to get away. But once the hobby has succeeded, it raises its prey to its beak while still flying, consuming as it goes. All that is left are the dragonfly wings which flutter down to the ground.

This amazing performer can only really be confused with the kestrel, although with a good look all doubt falls away. The hobby's wings are longer and narrower, held back like a swift's. Closer views reveal a blueish-grey back, prominent black moustache markings and strongly barred underparts.

All in all, a most wonderful sight. Unless, of course, you happen to be a dragonfly.

Hobbies can be found further north than a few years ago – sometimes even reaching Scotland – reflecting the similar expansion of many species of dragonfly.

Facts

Frequency
Arrives in April and stays until September or early October, with up to 1,000 pairs breeding here each year. The hobby can be found across eastern, southern and southwestern England, with a few making it into Wales.

Identification
At a distance it looks like a giant swift with its long tapering wings. Closer up, look for russet 'leggings', dark blue-grey upperparts and heavy barring underneath. Juveniles lack the colours, but are almost equally as fast and agile.

Song
A standard small raptor's call of *keekeekeekee*.

Nesting
Hobbies breed on farmland and heaths where occasional clumps of trees can be found. Sometimes they choose the edges of woodlands as nesting sites.

Length
30-36cm (12-14in)

Did you know?

■ Gravel pits are good places to look out for hobbies as they're ideal breeding grounds for many species of dragonfly.

■ Like the high-flying birds that it preys upon, the hobby migrates each year to and from sub-equatorial Africa.

House Martin
Delichon urbica

It's said that one swallow doesn't make a summer. A rider to that is that one house martin just might. A visitor from Africa during the longer months, the house martin tends to arrive a little later than its fellow migrant, normally around April, and leave a little later, possibly remaining even until October. Once it is here, it lives up to its name, and like the human population itself, goes on the hunt for a good house in which to live.

House martins are generally urban birds, making their homes in the eaves of buildings, from which they then swoop out and away to spend their day feeding on aerial insects (if you

do have house martins nesting in your house, check to see whether open windows might impede their flight path).

Interestingly enough, one of the martin's greatest rivals is the apparently innocuous house sparrow (again, note the name – it's all about houses, here). Sparrows do have a habit of breaking and entering martins' homes in search of a nest site of their own, sometimes even destroying the eggs in the process. You can deter the sparrows, though, by hanging a series of weighted strings about 15cm in front of your martin's nest. Martins have no difficulty in flying

Facts

Frequency

The house martin is found virtually everywhere in the country during the summer months, except for the northern tip of Scotland, and mainly in urban areas. As it feeds on aerial insects, it can often be seen during the day above water and on the fringes of woodland. Approximately 3-500,000 breeding pairs, but declining.

Identification

Stockier than the swallow, the house martin's forked tail is also much shorter. When seen from above, its white rump helps to distinguish it from both swallow and sand martin, although the latter's brown colour is also a good clue.

Song

A chatty twittering call.

Nesting

The birds build mudball nests under the guttering of buildings.

Length

12.5cm (5in)

The house martin has short white feathers on its feet, even on its toes.

Did you know?

■ House martins can often be seen resting on wires and buildings.

■ The bird often returns to its exact same nest site year upon year.

up to their nest at a steep angle, but sparrows do, and they should be deterred by the strings. (House sparrows do need encouragement to breed, though, so if you do deter them from the martins' nests, provide alternative nestboxes for them elsewhere in the eaves).

If you do have martins nesting in your home, take a good look at them. In the air, they appear to be predominantly black and white birds, but a close look through a pair of binoculars at a bird clinging to the outside of its nest reveals a plumage of rich velvet which, when it catches the light, shows you that summer has truly arrived.

'No jutty, freize, buttress nor coigne of vantage, But this bird hath made his pendant bed and procreant cradle'

From *Macbeth* by William Shakespeare

Graham Harrison, Warwickshire Wildlife Trust, advises:

■ House martins are traditionally cliff dwellers, but most have now adapted to nesting on buildings. Confiding birds, they readily build their mud-cup nests beneath the eaves of houses, often using an apex or projection for added support. Breeding even occurs in town and city centres and is usually in small colonies averaging around five pairs.

■ Having wintered in Africa, south of the Sahara, birds begin to arrive here from late-March, though it is often May before colonies are reoccupied. They then leave again in September and October. Aerial feeders, their cheery 'chirrp' call is a familiar sound as they hawk insects over our gardens; high in the sky during fine weather, but lower down in cool, damp conditions.

■ Sadly, numbers are declining, but you can help to prevent this. The mud needed to build nests is getting harder to find in our drier springs, so keep a small, muddy patch in your garden. Better still, put up special house martin nest boxes. These give the birds a head start and can enable them to raise two, or even three, broods. Moreover, nest-boxes can be sited where droppings will cause least nuisance. In return, the martins will reward you by helping to rid your garden of pests such as aphids.

House Sparrow
Passer domesticus

The gregarious house sparrow is a joy to have in any garden. They will nest in boxes, take food from bird tables and feeders, and by providing these things you are helping a once-common bird that has vanished from some areas.

House sparrows are one of the great opportunists. They have found a life for themselves in the countryside and in urban environments. But all is not well. Over the last 100 years, sparrow populations have gradually declined and in the 1990s, their population crashed. There has been much speculation about what has caused their spectacular demise: the lack of suitable nest sites; reduction in food supply; increases in predation; and the use of toxic additives in unleaded petrol have all been implicated.

House sparrows haven't declined dramatically everywhere and if you still have them in your garden or local area you can help. They nest in loose colonies, in holes in buildings such as under the eaves or behind the fascias and soffits of roofs. Sparrows will take to boxes if these other sites aren't available. For these colony nesting birds, put a number of boxes in close proximity (20-30cm apart) or use a special terrace box.

Facts

Frequency
The house sparrow has suffered heavily in recent years, but it's an indication of how abundant it was that it's still Britain's commonest garden bird. It can be found anywhere where there are people, from city centres to arable land. It is, however, rarely found in upland areas, including the Scottish Highlands.

Identification
A chestnut streaked back and pale underparts, and a nut-brown nape, grey crown, white cheek patch and black bib make up this familiar bird.

Song
The familiar, archetypal *cheep cheep*.

Nesting
Rarely found nesting in hedges or conifers, house sparrows prefer to nest in holes in buildings or nest boxes. They sometimes take over disused house martin nests. The hole is filled with straw or dry grass then lined with a variety of items from pigeon feathers to hair, and even string or paper.

Length
14cm (5.5in)

As it is a ground-feeder, one of the house sparrow's greatest enemies is the domestic cat.

One problem is a shortage of food, particularly seeds. In urban areas, development on brownfield sites and makeovers of gardens means fewer overgrown corners, reducing the quantity of 'weed' seeds available. In rural areas, efficiency in cereal harvesting has reduced the amount of grain spillage and tighter hygiene controls have sealed barns and silos against birds. In your garden, try providing food such as sunflower hearts and millet, or allow an area to be wild with annual plants that will provide seeds such as chickweed, fat hen, groundsel, shepherd's purse and vetches.

By providing nest sites and food for house sparrows, gardeners may play an important part in their conservation, perhaps preventing their disappearance from more areas of the country.

Did you know?

■ During the autumn, sparrows often leave gardens and congregate in large flocks in arable areas, seeking leftover grain from the harvest. By winter, however, they are back at their nesting sites, which they reoccupy throughout the colder months in preparation for next year's spring.

■ House sparrows are so adaptable that they've taken root in a host of other countries around the world where they've been introduced.

'Stop feathered bullies!
Peace angry birds,
You common Sparrows that,
For a few words,
Roll fighting in wet mud,
To shed each other's blood.'
From 'To Fighting Sparrows' by WH Davies

WHAT IF SOMEONE YOU HAVE NEVER SEEN OR NEVER HEARD OF IS THE ONLY ONE FOR YOU?

Just Woodland Friends Country Introductions could be the answer! (ABIA)

Successfully matching country-loving folk throughout the UK since 1992. Contacts sent by post or email. We like to give people the choice! Wet dogs and muddy boots neigh problem!

Enquiries & brochure Tel: 08453 70 81 80. Or see our website for further information where you can also join online.

www.justwoodlandfriends.com

Are Birds Pests?

Although some birds can be pests, most cause no damage in gardens and can actually be quite beneficial. Here, Andrew Halstead of the RHS presents the facts.

More than 140 species of birds have been recorded in British gardens, but only about 30 species are common garden residents. A few species cause problems, such as pigeons eating brassicas and peas, blackbirds taking fruits, herons eating fish and bullfinches eating buds on fruit trees. Plants can be protected with 12mm mesh, which should be fitted to a frame or cage so that the netting is taut and less likely to entangle birds.

The majority of birds are helpful in the garden. Some, such as sparrows and finches, feed mainly on seeds, including those of weeds. Others, such as tits, robins, wagtails, starlings and wrens, eat insects and other invertebrate animals. Song thrushes hammer snail shells against stones in order to get at the contents. A pair of blue tits with eight young will need to collect about 10,000 caterpillars, so birds can be significant predators of garden pests. Even seed-eating birds will collect insects for their chicks during the breeding season.

These pest-eaters can be provided with natural food by growing plants that will provide them with seeds and berries. Delay cutting down herbaceous plants with seed heads until late winter in order to give birds an opportunity to feed. Supplementary food can be provided for birds.

Be tolerant of insects in your garden. If pests still need controlling, try using pesticides based on natural materials, such as pyrethrum, derris/rotenone, fatty acids or plant oils. These have short persistence and low toxicity to birds.

MIKE CALVERT

Jackdaw
Corvus monedula

If you live anywhere near a busy jackdaw winter roost or breeding colony, you already know it. At dusk, the sky is darkened by a turbulent blizzard of black birds, banking and diving, tumbling and swerving to miss each other in the hectic traffic. It always looks as if they are being tossed about in a chaotic melée by the wind. In fact, they are just testing their aerial prowess in the breeze. And all the time these evening congregations are played out to a backing track of raucous *tchaak- tchaak* and *kyaw* calls.

Jackdaws are a common presence in many towns and villages, where you see them perched on roofs, circling church towers and foraging, mainly on the ground, in nearby gardens, parks and playing fields. Sometimes they'll perch on the back of sheep and cattle to pick off ticks. Mostly they delve for insects, including caterpillars, beetles and flies, as well as spiders, snails, slugs, earthworms and small frogs. If a jackdaw happens upon a mouse's or vole's nest concealed in the long grass, it may eat the baby rodents. During the breeding season, like other crows, jackdaws steal eggs and chicks from the nests of songbirds, seabirds, owls and kestrels. Over the winter jackdaws have to broaden their diet by eating berries and fruit, grain and wild seeds. Ever the opportunists, jackdaws will

Facts

Frequency
Common and widespread across most of lowland Britain; breeding population up to 430,000 pairs and rising.

Identification
Smallest and most animated black crow; head is distinguished by ash-grey cheeks and nape; on the ground, sprightly strutting and lively jumps with head held high help pick out the jackdaws in a mixed flock of crows; in the air, look for quick, deep flapping of short rounded wings, without the parted feather-fingers seen at the ends of rook and carrion crow wings.

Song
Extremely clamorous, *tchaak-ing* and *kyaw-ing* non-stop; when danger threatens, gives out a metallic screech – *karr-r-r karr-r-r* – to raise the alarm.

Nesting
Builds in cavities in derelict buildings, church towers, tree holes, on coastal and quarry cliffs, in chimneys and under bridges; assembles a higgledy-piggledy pile of twigs, heaps dried grasses and leaves on top and lines central cup with fine grass, hair and wool.

Length
33-34cm (13-13.5in)

The jackdaw is doing better than most; numbers have increased by 50% since the 1960s, thanks largely to its adaptability – being a bird of many talents, and a jack-of-all-trades when it comes to eating a variety of foods and nesting in different locations, has stood it in good stead in a rapidly changing world.

scavenge for scraps at garden feeding stations, waste tips and picnic sites and happily pilfer corn and grains from feeding troughs in chicken runs and pig pens.

Given half a chance in urban areas, jackdaws nest in chimneys: there are usually plenty of these vertical nesting holes in a street for establishing a nesting colony. To convert the flues into suitable nesting holes, the wily jackdaw drops twigs down the chimney. Eventually, a twig gets wedged and traps others that pile up on top to form a platform on which the jackdaws build their nest. Huge piles of kindling can block the chimney and become a fire hazard.

Did you know?

■ For a fascinating insight into the social life and complex hierarchy in a jackdaw colony, it's worth reading chapter 11 of *King Solomon's Ring* written by the pioneering animal behaviourist Konrad Lorenz. He noted that a male usually picks a lower status female as his mate, whereupon she is immediately elevated to his position in the group. When one member of a flock is injured or in danger, others rally around to help the injured party to feed and unite to attack intruders, although they have been known to turn on weak or sick members of their own group and mob them to death. They are also most hostile to fluttery black enemies, in the form of other nest-robbing crows.

■ Look closely and you will see that adult jackdaws have startlingly pale blue eyes, while youngsters have brown eyes.

'The clamorous daws that all the day Above the trees and towers play.'

From 'The Eve of St Mark' by John Keats

Jay
Garrulus glandarius

It sometimes seems that there are two types of British jay. Those found in woodlands tend to be extremely secretive birds, keeping a watchful eye on you as you approach, and flying away with a raucous *krark* before you've come anywhere near them. Their urban brothers and sisters, however, seem to have become a little more accustomed to the human presence, and the jays that frequent gardens are a good sight braver and more confident. They'll come to

bird tables, have a go at your peas and beans, and even strut about on your patio if there's something of interest there.

It all comes down to adaptation. Those birds that live in rural areas have no need to alter their habits, but those that live on the fringe of suburbia have to change, otherwise the march of urbanisation will swallow them up. But then jays are perfectly placed to change. They don't particularly look like it, but jays are members of

A jay can carry up to nine acorns at one time – one in its beak and the rest in its extremely large throat pouch.

the crow family, and crows are among the most adaptable birds of all.

All you have to do is listen to them to tell. Jays are good mimics, able to replicate the sound of an angrily mewing cat, which they use to chase away the real thing, as well as other sounds such as the calls of other birds. Mimicry is a good indication of intelligence.

The bird's intelligence stretches to other areas, too, particularly in the flexibility of its eating habits. During the breeding season, jays become good mousers, killing their prey with a few well placed blows to the head. In the autumn they feed on nuts and acorns, storing many away for the cold winter months when they need them. Unlike squirrels, who can rarely remember where they hid their nuts, jays have excellent recall, and can even find their secret stashes under thick coverings of snow.

One sight you may see in your garden is of a jay apparently in distress, with its head bowed, wings out, tail spread, and its body covered with insects. Don't be alarmed. The bird is merely cleaning itself by anting. Squatting over an anthill, it allows the insects to crawl all over its body, which they cover with their noxious formic acid. The acid causes the bird no harm, but it does deal efficiently with any ticks or mites that the jay might have, and is an excellent fabric conditioner for feathers.

Facts

Frequency
England and Wales are the stronghold of the jay, which is a common bird in lowland areas, but much scarcer in upland areas where trees are at a premium.

Identification
A body covered in a dark pinkish blush, chestnut wings with a bright blue and black check pattern, a streaked crown and a white rump patch in flight make the jay very distinctive. Both sexes look the same, and while young birds are similar, their colours are slightly duller.

Song
The jay has a host of calls, many of them expert impersonations of other creatures. Staple, though, are the harsh *krark, krark* of alarm, and the soft gurgle that accompanies courtship.

Nesting
Nests built in trees and bushes are formed by twigs bound with mud and lined with horsehair.

Length
34cm (13in)

Did you know?

■ Jays were once killed for their beautiful blue wing feathers, which were used in millinery and to make flies for fishing.

■ Fervent hoarders of nuts, particularly acorns, jays build up stores that can last them for a full year.

■ Jays form pair bonds for life, and these partnerships are called 'marriages'.

'A jay, screeching hideously, made off with guilty flight and a glitter of blue-enamelled wings.'
From *Birds of the Grey Wind* by Edward A Armstrong, an Irish parson/ornithologist

Kestrel
Falco tinnunculus

You may never actually see a kestrel in your garden, but there's a very good chance that you'll see one from it. Kestrels are excellent hunters and have adapted superbly to the urban environment, most famously hanging by the sides of motorways watching the ground below for tiny movements of rodents disturbed by the roar and vibrations of the traffic alongside. They're comfortable in towns, too, and if you have a church spire or other high vantage point nearby there's a likelihood that it's a perching post for a kestrel as it examines the world below for opportunities. Parks, docks and railway sidings have all been discovered by the ever watchful kestrel, too.

If you do see a kestrel hovering above your garden, watch it closely. For a start, it'll tell you which way the wind is blowing. Kestrels maintain their position by facing the wind and flying into it at exactly the same speed that the air is moving, a feat of remarkable physical precision. They keep their heads rock-steady, and can spot a beetle at 50 metres and a small bird or rodent at 300m. Kestrels are also aided by their ability to see ultra-violet light. Voles, one of the birds' main prey, keep to fairly regular tracks as they forage which they mark with urine for territorial and courtship reasons. The urine reflects UV light, enabling the kestrel to map out the small animals' routes.

When the female needs to take a feeding break, the male kestrel will often take over the incubation of the eggs.

Facts

Frequency
A little over half a million pairs breed in Britain and Ireland each year, although these numbers are declining. The bird can be found throughout many British habitats.

Identification
The male has a blue-grey head and rump with a russety brown back and backs flecked with black chevrons. The same markings appear on its underparts, which are otherwise cream coloured. The female lacks the colours of the male, but is bigger. The main bird of prey that the kestrel is likely to be confused with is the sparrowhawk, although if you look closely at the two birds' silhouettes as they fly above, you'll notice that the kestrel has much more pointed wings.

Song
Other than a series of noisy *keekee* calls during the breeding season, the kestrel has very little to say for itself.

Nesting
This is a bird of adaptability. Old nests, such as those belonging to crows or squirrels, are readily used by kestrels, as are tree holes, building ledges, church towers and even the top of Nelson's Column in London.

Length
32-35cm (12.5-14in)

In the winter, however, when food is scarcer, to conserve energy the kestrel will tend to hunt from a suitable vantage point, such as a telephone pole or street light. Wheelie bin day is a good day to look out for them, as the movement of rubbish can help get rodents on the move.

Kestrel numbers, however, are in decline. Changes in agricultural practices have severely reduced their source of invertebrate food, and although the bird has adapted very well to the urban environment, it still struggles with the hurly-burly pace. Many birds are killed by traffic too, particularly alongside motorways.

Yet it's a popular bird, and many people are working hard to protect it. One ingenious scheme involves fixing nest boxes to motorway bridges. Because the birds try to nest under the bridges, their young often die due to the constant flow of traffic and through draught that knocks them out of the nest, but boxes on the sides of bridges provide a highly viable alternative.

**'Thou dost not fly, thou art not perched
The air is all around:
What is it that can keep thee set
From falling to the ground?'**

From 'The Hawk' by WH Davies

Did you know?

■ The film *Kes* actually led to a slight drop in kestrel populations, as many small boys stole the birds' eggs in an attempt to emulate the film's young hero.

■ During the reign of Louis XIII of France, kestrels were trapped and trained to catch bats.

Linnet
Carduelis cannabina

Ever noticed that the word linnet sounds a little like linen? It's no coincidence. Both words come from the same source – *Linum*, the Latin for 'flax'. Flax is the source of linen fibres, and it was once the linnet's favourite food.

Flax is no longer in plentiful supply, but the linnet has adapted. The seeds of brassicas, such as chickweed, charlock and mustards, as well as dandelions, have become adequate replacements. If seeds are scarce, or herbicides have been used, this small finch often moves on to strawberry farms, where it settles under cloches and picks out the seeds from the fruit.

If you live near open land, you can encourage linnets into your garden with dandelions, thistles, dock and plantain. It's worth leaving a patch of these plants for them, as the linnet has a delightful, metallic tinkly song, and large flocks of them even sing together in chorus. This happy-sounding melody has not served the bird well in the past, though, and the linnet was one of the most popular choices of caged bird in the 19th century, when bottling up songbirds to provide musical entertainment was all the rage. The duetting nature of a linnet couple also made them a symbol of love, and

The second part of the linnet's scientific name – *cannabina* – derives from the bird's habit of eating hemp seeds on the Continent.

Facts

Frequency
Although the linnet is still widespread and common in lowland Britain and Ireland, numbers are falling due to continued herbicide use.

Identification
At a distance, linnets look like rather anonymous little finches, but closer up the detail really shines through. The male has a ruddy back, and in summer sports a grey head with crimson crown and a blood-red blush on its breast. Females lack the colour, though. In flight, look out for white edges to the wings, and a dark slightly forked tail.

Song
This has been likened to a gentle clattering of metal teaspoons, and it bubbles along as an entire flock of linnets sings together. The flight call is a gossipy *chichichichit*.

Nesting
The linnet nests close to the ground, often in a gorse bush or hedge, and lines its rough nests with horsehair, fluff and wool. When the nest is threatened, the mother will flap off and pretend to be injured, in an attempt to draw the intruder away.

Length
13.5cm (5in)

Did you know?

■ Although it is illegal to keep wild birds in captivity, there is still an underground trade in capturing and caging linnets and selling them in Europe.

■ Linnets are incredible parents. After one heath fire, a mother was found burnt to death still sitting on her nest – and, remarkably, her chicks were still alive underneath her.

■ Linnets are so seed-based in their diet that even their young are fed crushed seeds. This is unusual for nestlings, which are normally fed on insect life for the first weeks of their life.

'Upon this leafy bush
With thorns and roses in it
Flutters a thing of light
A twittering linnet'
From 'The Linnet' by Walter de la Mare

many a courting young man would present his intended with a pair. So great was the enthusiasm for linnet ownership, in fact, that many people even made a living from picking plants that provided food for them. In the mid-19th century, some 15 million bunches per year were sold at market venues such as Covent Garden.

So if you live near heaths with gorse bushes, open commons, scrubland or farms, listen out for the sweet song of the linnet, and enjoy the fact that today it is a song of freedom.

Long-tailed Tit
Aegithalos caudatus

On a grey winter's day, a surprise visitation from a small flock of dainty long-tailed tits instantly brings a garden to life. Heralded by a high-pitched fanfare of squeaky calls, they suddenly breeze in and the bare branches of the trees come alive with tiny birds, trailing extraordinarily long tails like ribbons behind them. They somersault around the finest twigs, flit from bough to bough, all intent on finding any insects and spiders living in the bark and among clumps of moss and lichen, or swarm over the peanut feeder. No sooner have they arrived, than they're moving on to the next refuelling stop.

Although each long-tailed tit appears to be mapping out its own route through the branches, there is a strong camaraderie between members of the flock. If one becomes detached from the party, it gets agitated and anxiously calls out to its companions, while the rest stop and look for it until they're reunited. Togetherness is the key to long-tailed tits' survival, especially during cold weather.

Members of a flock depend on being able to cuddle up to one another to keep warm enough to endure a freezing winter's night.

Most of the time, long-tailed tits live in small family flocks, comprised of the parents and their offspring of the previous breeding season. For the winter, several family flocks temporarily unite to form groups of 300 birds or more. In early spring, the flocks disband and unmated females join neighbouring groups. At breeding time, long-tailed tits take family loyalty and cooperation to exceptional lengths. If a pair fails to raise its first brood, it'll have missed out on the early-season glut of insects essential to feed them. Often, the two birds split up and the female usually returns to her family flock. Apparently, she can distinguish her clan members from others by their calls. Then she helps to rear the young in other nests belonging to her brothers.

Such behaviour is not as altruistic as it sounds. Having failed to breed themselves, by assisting relatives to raise their large families,

**'And coy bumbarrels twenty in a drove
Flit down the hedgerows in the
frozen plain
And hang on little twigs and
start again.'**

From 'Emmonsails Heath in Winter' by John Clare
('bumbarrel' was a former name for the long-tailed tit)

Facts

Frequency
Numbers fluctuate depending on how many survive the winter; currently about 273,000 breeding pairs.

Identification
Tiny tousled pompom-like body with over-long tail, red eye rings and very small black beak; from afar, the plumage can look black and white, but at close range there's a pinkish wash over the underparts and shoulders; juvenile has no pink in its plumage.

Song
Rich medley of shrill *tsee-tsee-tsee* and *tsirrup* contact calls.

Nesting
One of the great wonders of bird architecture: both birds build a small, well-camouflaged globe of moss, dry grass, wool and hair, covered in lichen and lined with hundreds of feathers, in prickly gorse, blackthorn, brambles or hawthorn.

Length
13-16cm (5-6in), including a 6-10cm (2-4in) tail

---

helper aunts and uncles are ensuring that their family genes are passed on to the next generation. Chicks reared by an extended family get more to eat, are heavier on fledging and survive better. And importantly, the helpful brothers and sisters also book their potentially lifesaving places in a flock for the winter. Long-tailed tits have prospered in the recent run of milder winters, although wintry cold snaps can take a heavy toll; their relatively recent discovery of garden freebies has eased the pressure on finding enough to eat: a well-stocked peanut feeder may be a lifesaver.

Their unusual pinky colouring is just one hint that, despite their common name, long-tailed tits are not true tits, like the blue tit, great tit, coal and marsh tits. In fact, if their calls, nesting behaviour and flock dynamics are anything to go by, they're more closely allied to some sociable Asian birds called babblers.

Did you know?

■ Long-tailed tits build one of the most astonishingly intricate nests of any British bird. After picking out a nest site in the depths of a prickly bush, a pair starts weaving moss, cobwebs, wool and hair into a strong matted base. Then, working from inside, they build up thin elastic walls and a domed roof, leaving a small entrance hole near the top. The next step is to peel up to 3,000 flakes of grey-green lichen from tree trunks and clad the outside of the walls to make them weatherproof and well-camouflaged. Finally, they stuff the globe with around 1,500 feathers to form an insulated lining – the most feathers ever counted on the lining of a long-tailed tit's nest was a staggering 2,680. After 30 days and flying hundreds of miles to collect materials, the nest is ready.

■ It's not hard to work out how the long-tailed tit got its common name. Indeed, it can proudly claim to be the bearer of the longest tail of any British bird, when compared to its body size. Without its 8cm (3in) tail, the long-tailed tit would be just a tiddly ball of fluffy feathers, only 6cm (2in) long.

Helen Bostock of the RHS explains why water is such an important component of garden life for birds.

Even if you can't provide food for birds, try to put out a supply of clean water. The smallest of courtyard gardens or balconies can provide space for a shallow bowl. Keep water available at all times of year. Small quantities can quickly be used up and evaporate in summer, so ensure it is topped up. Equally, in winter the water may freeze over. For small containers, melt the ice with some warm water or break up the frozen layer. Don't crack the ice on ponds, however, as this disturbs pond life. Instead, sit a bowl of hot water on the ice to melt a hole. Liquid products and insulated water containers are available to keep bird baths

ice free. Anti-freeze products for other uses are not suitable.

Cover water butts when not in use. The sight of water can tempt birds and other wildlife to enter, only to find no way out. Straight-sided water features can also lead to birds drowning, especially if there is an overhang. Make these safe by adding wooden ramps, wrapped with chicken wire, or design ponds with shallow sides. Birds will also use logs or stones partially submerged in the pond as a perch from which to drink.

Moving water appears to be particularly attractive to birds. Battery-powered agitators can be placed in the water to create a ripple effect. Alternatively, bubble

Fact file

Dust bathing

Not all bathing takes place in water. Some species, like house sparrows, also like to 'wash' themselves in earth or dust, splashing around just as if they were in a bath. The probable reason for this is that the earth helps them to remove old traces of preening oil that has built up in their plumage and that water simply washes over.

Drinking techniques

Most birds drink by scooping up beakfuls of water, tipping back their heads and swallowing. Woodpigeons and collared doves, however, can drink as mammals do, simply lowering its beak into the water and sucking it up.

Safe place

Birds are not the only creatures drawn to bird baths: cats soon realise that they can provide easy offerings. Make sure you site yours where birds can see potential danger, and ideally close enough to cover for them to escape to it.

fountains and natural or artificial streams are ideal.

Avian diseases are more readily spread at communal watering sites, so use a stiff brush to scrub out bird baths and change the water regularly, daily if possible. Watch out for scruffy, listless birds that might indicate an infection. Use only mild detergents for cleaning bird baths. There is no need to use water purifying agents, provided the water is changed regularly.

Water in the garden is, of course, used by birds for bathing as well as drinking. This is critical for feathers to remain in good condition, both for insulation and flight. It is therefore important to provide a bathing area in winter and in summer. Birds prefer to bathe in shallow water, splashing about as they do so. Puddles formed by rain in uneven paths and drives are also used for bathing, but do be careful as these areas can be slippery.

Garden birds will enjoy a dust bath in summer to help eradicate external parasites. Simply keeping a patch of bare, sandy soil in full sun that will dry to a suitable consistency is all that is needed.

Bird baths

Nick Brown, Derbyshire Wildlife Trust, advises:

■ There is something very entertaining about watching birds bathe and drink. Bathing especially is fun to watch, if not somehow a little prurient! Some birds (woodpigeons come to mind) get so engrossed that they go into an ecstatic state, lying there, wet through and staring skywards!

■ Bathing is something that birds must do to keep their feathers in good shape and free from dust and parasites. If you provide them with suitable bird baths some species, such as blackbirds, woodpigeons and robins, will bathe regularly: Sometimes, often surprisingly, during or after rain, birds suddenly feel the desire to bathe and do so one after another, or in the case of starlings, all at the same time! Other species bathe rarely or not at all.

■ Some, such as the chiffchaff, seem to prefer to take a leaf-bath, rubbing their feathers against wet foliage, presumably reducing their susceptibility to predators. Others, such as sparrows, prefer a dust bath.

■ Drinking is another essential activity for birds, especially seed-eaters like greenfinches. You might expect to see most of the birds that visit your garden also take a drink although, as with bathing, some species prefer to use puddles or leaves. In dry weather insects, such as wasps, will come for a drink, as will small mammals, hedgehogs and foxes.

So, bird baths are essential to any des res wildlife garden – and the more the merrier. Here are a few pointers:

■ Provide both shallow and deep bird baths.

■ Put some low down, close to cover, and others out in the open.

■ Of course, if you have a pond with a shallow margin, birds may use it for drinking or bathing, but a decent bird bath usually attracts more species, providing them with a flat substrate and a known depth of water.

■ There are many fancy designs available, though I still use an upturned plastic dustbin lid and ceramic planter bases.

■ A final reminder. Never use anti-freezing agents. The best plan is to empty baths before nightfall, and then refill them in the morning with tepid water. Hot water either cracks the bath or, by some strange quirk of science, freezes over more quickly than cold!

Ponds

Ponds are great way of encouraging birds to your garden, but sadly during the past 100 years, the UK countryside has lost almost 70% of its ponds, resulting in the critically damaging reduction of a major form of habitat. The creation of new ponds has never been more important for wildlife, and here, Helen Bostock of the RHS explains how to put one into your own garden, and really make a difference.

There is a wealth of wildlife that depends heavily on ponds for its success. Most of it will not only make a welcome addition to the diversity of your garden, but will be of great benefit to it, too. Small inhabitants of your pond will be numerous and will include pond skaters, water beetles, snails, mayflies, caddis flies, damselflies and dragonflies. Where there are insects, amphibians are not far behind. Frogs, toads and newts are all quite happy breeding and living in small bodies of water, as long as it's deep enough – around 60cm (24in) – for their purposes. Several bird species will enjoy drinking and feeding from your pond, too, and if it's large enough, you might attract swallows and house martins that swoop across the water's surface, plucking off insects, and using its mud to build their nests.

What type of pond works best?

Wildlife doesn't make any distinction between natural and man-made ponds, shape being far more important. Try to build in a long, shallow slope on at least one side of your pond; it will allow easy access for wildlife into and out of the water, and creates a vital damp habitat for beetles, bugs and flies. If space is at a premium and steep sides cannot be avoided, place a stone or wooden ramp in one corner to help amphibians – or unlucky hedgehogs that have lost their footing – to find a way out. Try to minimise fish stocks, which keep wildlife levels down.

How big should it be?

No matter what size your pond is, it will attract wildlife – although, of course, the larger you can make it the better. If you can vary the depth across the pond you will suit a good variety of plants and creatures. Make sure, too, that you have a shallow end for amphibians to crawl in and out. The pond should be in a sunny spot with shade over part of it. This helps keep down algae and is tolerate by many plants and animals.

What should I plant?

You don't have to plant up your pond at all. Natural colonisation by plants and wildlife usually happens quicker than expected, though may take longer in sites isolated from other ponds. Most people, however, prefer to control the look of their ponds.

Marginal plantings provide important areas of cover, and plant stems at the water's edge are needed for emerging damselfly and dragonfly nymphs. Aim to achieve 65-75% surface coverage with floating aquatics. Some submerged planting (often called oxygenators) is equally important. Use native plants where possible and avoid known invasives, such as fairy fern (*Azolla filiculoides*), New Zealand pygmy weed (*Crassula helmsii*), parrot's feather (*Myriophyllum aquaticum*) and floating pennywort (*Hydrocotyle ranunculoides*). Don't forget that you'll need landowners' permission if you're collecting plants from local ponds and ditches. Dead branches in the pond will enrich the habitat, as do tree roots growing into the pond. Resist removing overhanging branches that naturally dip or fall in the water.

How often should I top up the pond?

Don't be too hasty to top up the pond during dry weather in late summer. Seasonal ponds are a natural feature in the UK, filling up in winter and occasionally drying out in hot summers. This seemingly inhospitable environment can favour certain animals. Newts, for example, are able to survive in the mud during dry months, unlike fish who predate on newt larvae. Where additional water needs to be added, try to use rainwater. Tap water should be a last resort.

Is silting up a bad thing?

The natural progression of a pond is to fill in until it becomes wetland and each stage of this process has its own unique wildlife. If you wish to remove sediment to maintain your pond, try to take away only half at a time to minimise the loss of mud-dwelling creatures and their habitat. There is no ideal time to do this, though late summer when the water is naturally at its lowest is the most practical period.

How can I find out more?

More on pond-building and maintenance can be found on the following page. For even more information, try to get hold of *The Pond Book: A Guide to the Management and Creation of Ponds* (published by The Ponds Conservation Trust, 1999), or visit their website www.pondstrust.org.uk. Whatever type of pond you decide to create, make sure you enjoy it. The wildlife certainly will.

149

Before you start

■ The size of a pond and its position within the garden needs to be assessed to obtain a balanced and harmonious relationship with other garden features. Try to choose a site that, at least in part, benefits from the sun, to obtain maximum value from reflections of sun and sky, and from water lilies, whose flowers only open in direct sunlight. Position carefully in relation to deciduous trees, whose leaves may otherwise fill the pond in autumn.

■ Algae will form in the initial stages, but once the balance of the pond is established, it should cease to be a problem. In practice, a small, shallow pond soon becomes algae-infested or overcrowded. It is therefore important to have an area of deeper water in small ponds to reduce such problems.

■ The easiest way to mark out any proposed site is to lay rope to form the outline. Mark the area of any associated plantings, then look at the site from all angles. When satisfied, draw a rough plan of the site, then cut around the rope marker with a spade, before removing the rope. Avoid extreme shapes. Also avoid precise geometrical shapes unless the surroundings are formal.

Helen Bostock, RHS, advises:

Royal Horticultural Society

■ Even shallow containers make useful ponds. Ensure there is some form of ramp to allow creatures access. With some water plants, the water should remain reasonably wholesome. Be aware that even shallow water can pose a threat to young children. A shallow bog is safer than a pond.

■ Concrete is the most durable pond-lining material when soundly constructed, and is particularly effective for ponds in formal settings. Paving stones cemented on top of the vertical walls make a stable and attractive surround. Although concrete ponds are easy to clean, if cracks develop, they can be difficult to repair. Concrete must be treated or allowed to season before introducing fish or plants.

■ When building a concrete pond, excavate the pond area, allowing an extra 15cm (6in) all round and at the base, to allow for the thickness of the concrete. Walls angled outwards by 20

degrees from the vertical are less vulnerable to winter ice pressure, but shuttering in this case is more difficult. Remember, that a shallow side is needed for access by wildlife. If you are not familiar with shuttering and concrete, check DIY publications.

■ For larger ponds, the concrete may need reinforcing and expert help is advisable. Plan construction so the actual laying of the concrete floor and the pouring of the concrete into the wall shuttering is completed in one day, to ensure that the finished pool is waterproof.

■ If concreting in winter, cover all concrete surfaces against frost for four days. In summer, or when very warm and dry, water the concrete soaking shutterings and concrete, each day. The setting of concrete is a chemical reaction and the slower it sets, the harder it will be and the more resistant to cracking. The supports can be removed two days after concreting, and the shuttering from four days afterwards. Soften sharp edges of concrete with a trowel, then sweep out the pond to clear it of concrete and cement dust, and fill with water. Leave the water for two days, then empty the pond.

Lined pools

Plastic liners are variable in longevity, and dependent on quality and cost. 500 gauge black polythene could be used but may need replacement after as little as three or four years. PVC liners, single, double-layer or reinforced are more durable. A drawback is that exposed margins degrade and deteriorate from the effects of the sun's ultra-violet radiation. The best safeguards are to keep pools well filled and to plant masking marginal plants, particularly along the south-facing edge of the pool. Repair kits are available if accidental damage occurs but repairs can be difficult to carry out.

Butyl is a rubber material, heavier and more expensive than PVC but easy to install and with a life of 25-50 years or more. It can be successfully patched if accidentally pierced. It is available in black only, the most suitable and natural colour for a pool liner.

To calculate the size of liner required, measure the maximum length and width of the marked-out area. To each measurement add twice the depth, then allow an overlap of at least 15cm all round so the liner can be held firmly by paving or tucked under turf. The liner can be laid directly onto the smooth, stone-free excavation, with its slightly sloping sides. Draw it over the hole, positioning it and carefully hold it in position with bricks. Then let water from a hose gradually weigh down the liner into the hole, smoothly and with a minimum of creasing or wrinkling.

Commercially available underfelts should be used beneath liners. Fibreglass roll, as used for loft insulation, can also be used. Pools where the floor is flat and there are no sharp obstructions, can be stepped into with reasonable care, and even have stone rock-work laid in them to continue a rock garden. Set any rocks or statues on polythene or butyl offcuts to make future removal possible without damaging the liner.

Preformed pools

Cast in various shapes and sizes from plastic or fibreglass, these are rigid, lightweight and easily installed. Fibreglass is very durable but the edges of rigid liners can be difficult to mask. The exposed edges of plastic models can become brittle and crack with age. Choose darker internal colouring, as it is more natural in appearance and more aesthetically pleasing. Few preformed ponds are a good shape for wildlife, lacking gently sloping sides. A flexible liner will always give better scope for designing a pond with wildlife in mind.

Leigh Hunt, RHS, advises:

Royal Horticultural Society

■ Using stored rainwater is the best advice for filling ponds, but what do you do when the rain is so sparse that the rest of your garden becomes short of water? In periods of drought when there are bans on the use of domestic water supplies for garden watering, it is often suggested that domestic waste or used water, known as 'grey water', may be utilised to avoid plant loss. The most suitable waste water would be that in which there are few or no additives and no bleaching agents, such as water used for washing and preparing vegetables.

Magpie
Pica pica

You'd be forgiven for thinking that the magpie is one of the most common British garden birds, but in reality it doesn't even make the top 10. Yet while chaffinches, great tits and dunnocks flit in and out of the nation's shrubbery, often unnoticed, the magpie announces its presence with a grating call and wearing a showy suit, so that no-one can miss its coming and going. No wonder it seems such a frequent visitor.

Aggressive, noisy and undeniably big, the magpie tends to dominate proceedings wherever it goes. Few smaller birds brave the bird table when the magpie is around, but en masse they might give the big fellow a bit of a mobbing – after all, magpies are famed for plucking eggs and nestlings from their parents during the breeding season.

It's true that magpies have few friends, but they are certainly something to look at. As they strut across the lawn, or balance on a treetop, splaying their glamorous tail feathers for balance, they cut quite a striking figure.

> Magpies can sometimes be seen on the backs of sheep, plucking out tufts of wool for their nests. They perform a service to the sheep, though, ridding them of ticks in the process.

Facts

Frequency
Abundant and very common. The magpie can be found in both urban and rural areas across the country, although less frequently in northern Scotland.

Identification
A very striking bird, pied black and white with a long tail, which is an iridescent purpley green.

Song
You won't hear a magpie sing, but you will hear its raucous *chak-chak-chak* all year round.

Nesting
Look up and you're bound to see a magpie's nest. A large dome of twigs near the tops of trees, it's an untidy affair.

Length
46cm (18in)

Did you know?

■ Young magpies look very much like their parents, except they have much shorter tails.

■ Magpies are hoarders, putting nuts, berries and even carrion aside in the winter. They usually return to their stores within a couple of days, however, which they can easily find using their keen sense of smell.

One's sorrow, two's mirth
Three's a wedding, four's a birth
Five's a christening, six a dearth
Seven's heaven, eight is hell
And nine's the devil his ane sel'

Traditional

Pete Mella, Sheffield Wildlife Trust, advises:

■ No bird divides opinion quite like the magpie. A subject of superstition and folklore for centuries, it is now hated by many for a different set of reasons.

■ Its ubiquity, while the numbers of other birds seem to fall, has led many – including some corners of the media – to deduce that magpies are directly responsible for a fall in songbird populations. But, while magpies do have some habits we may find unsavoury – including the taking of newly-hatched birds – there's no evidence to support this theory; in fact the few areas where magpies are actually becoming more numerous are also seeing rising numbers of birds such as blackbirds and song thrushes.

■ Its large size and boisterous nature may not make it the most welcome of garden visitors, but the magpie is a highly intelligent and resourceful animal that can be interesting to watch. Its lack of fear of humans, and its opportunistic taste in food, mean it can be found in most gardens at all times of the year. Look out for its large nest of twigs in trees, which is usually domed to prevent predation from carrion crows.

■ Once prejudices are stripped away, the magpie is a beautiful bird; it perhaps deserves a second look.

Mallard
Anas platyrhynchos

Large gardens with good-sized ponds are most likely to play host to mallards, but even smaller gardens may receive the odd visit. Strangely, your visitor is most likely to be a duckling. If you live near a village pond or gravel pit, you may look out of your window one day and notice a little ball of yellow fluff waddling its way across your lawn. Because duck mortality is so high, mallards have large broods, often as early as March, numbering up to 12 young. With so many to look after, the occasional duckling slips away and becomes temporarily lost.

The key word here is 'temporarily'. Should you find a duckling wandering about by itself, the likelihood is that a parent is somewhere nearby, so it's best to leave it alone. Female mallards are extremely well camouflaged when they keep low among the leaves, so it could be that you simply can't see her, and that the two will soon catch up with each other. But if a long time passes and you can't hear the parent nearby, then the young bird might really be lost. The RSPCA (call 0870 333 5999) or the Swan Sanctuary (call 01932 240790) can give advice, and even take the bird in if they're convinced it become permanently separated from its parent.

By 'parents', we should say 'mother'. The drake mallard takes no part in the upbringing of his young, and exhibits all the attributes of wayward male. He'll watch carefully until the

'Among the tawny, tasselled reed
The ducks and ducklings float and feed
With head oft dabbing in the flood
They fish all day the weedy mud.'

From 'The Fens' by John Clare

Facts

Frequency

The mallard is common and widespread across Britain and Ireland and can be found on stretches of freshwater, from the fringes of the largest lakes to the tiniest ponds.

Identification

The male is unmistakable with his shiny dark-green head, purplish breast and white collar. The female is a mottled brown all over, but help in identifying many ducks comes in the colour of their wing patches, or speculums. In the case of the mallard, it's a rich violet.

Song

An easy one this – *quack-quack*! The male's call is slightly throatier than the female's, while ducklings simply *peep*.

Nestlng

The mallard is a ground-nesting duck, the female scraping out a slight hollow in dense vegetation, then lining it with leaves and her own feathers. She is superbly camouflaged.

Length

58cm (23in)

Did you know?

■ The mallard duckling is up and about within just hours of hatching.

■ The majority of breeds of domestic duck, including the Aylesbury, the Khaki Campbell and the Indian Runner, are descended from the wild mallard.

■ A mallard has fine comb-like 'teeth' which it uses as a filter to trap small food particles.

In the early stage of the moult, the drake loses his flight feathers, so he keeps as well hidden as possible.

gs are laid to make sure no other male mates ith his partner in the meantime, but once s progeny are settled in the nest he'll wander in search of another female. By the time the eeding season is over, he'll be in moult and, ith his feathers looking temporarily lacklustre, 'll slope off and leave his various partners to t on with bringing up his families.

Mallards are among the dabblers of the duck orld. Unlike tufted ducks or pochards, they not dive, preferring to stay at the shallow d of ponds and upending themselves to search food at the bottom. Added to their extreme undance, this makes them the most likely itish duck to use shallow garden ponds.

Marsh Tit
Parus palustris

A little over a century ago, it was believed that numbers of this bird were bigger than they actually were. It was only as the 19th century drew to a close that it was realised that the marsh tit and willow tit were actually separate species – these very similar little fellows having fooled the bird world for so long.

An explanation of how to separate the two appears in the chapter on the willow tit but, interestingly enough, your garden provides one of the best ways of separating them. If you've got one at your feeder, it's most likely to be a marsh tit. These little birds are not as shy as

Although marsh tits do like damp areas, they are most often associated with deciduous forests, not marshy areas as their name suggests.

Did you know?

■ The marsh tit is more likely than other tits to drop to the ground in search of insects and seeds to feed on.

■ It's a great hoarder and forager of food and, unlike most tits, if it finds a plentiful supply it will tuck its seeds behind bark or other hidden places.

Facts

Frequency

The marsh tit has been placed on the RSPB's Red Conservation List because numbers in Britain have declined since the 1960s – once their natural scrubland was cleared. Nonetheless, there are still some 60,000 breeding pairs distributed widely throughout the country.

Identification

A fuller description can be found in the chapter on the willow tit in this book, but the main feature to look out for is the black cap, which is shinier than its woodland neighbour's. Brown above, and pale on the belly, both the marsh tit and the willow tit lack the distinctive white nape of the coal tit.

Song

The marsh tit's call sounds like a sneeze – *pitchoo*. Bless you! It also has a scolding *chip-chip-chip* call.

Nesting

May be found at roost in tree or ivy (look for white droppings below). The marsh tit nests in hollow trees or large boxes.

Length

12cm (4.5in)

eir near-identical cousins, and can often be en at peanut feeders and will even use nest xes in your garden, particularly if you live ar woodland, their natural habitat. One of the st ways of enticing them in is to put out black nflower seeds, a particular favourite.

If you are able to encourage them in, they're ely to stay, as marsh tits are not as roving as her members of the tit family. A territorial ecies, they remain in their own locale with eir small family. This tree-hole-nesting bird ds to raise just a single brood per year, ually in the trunk of an alder or birch tree.

The mistle thrush is a bulky bird, larger and greyer than a song thrush. Some say it's untidier, too, as its colouring looks a little unfinished compared to the sleek lines and neat, clean plumage of its cousin.

Its flight pattern is another good indicator of which of the two species it is, as it tends to whisk along at about treetop height, beating its wings strongly, then settles for a brief spell into a glide, rattling out its harsh call (rather like the fieldfare, a winter visitor).

This thrush is fairly aggressive, defending its territory with vigour and even occasionally preying on other species' nestlings. Insects apart, however, carnivorism is rare, with fruit and berries being the main items on the menu. Unsurprisingly, given its name, mistletoe is a favourite, so if you've got tall trees in or near your garden that are hugged by mistletoe, there's a good chance that you'll be paid a visit.

This isn't a particularly migratory species, although some Scottish birds may over-winter in Ireland, or possibly even France. Similarly, there are a few of our winter mistle thrushes who spend the summer months in their breeding lands of Scandinavia.

These are very bold birds, who will form small flocks to defend 'their' berry bushes from other birds.

Facts

Frequency
Numbers are thought to be increasing, but this species is on the Amber List, which means there is some concern about its population. The mistle thrush is a resident, so is seen throughout the year. It's a regular visitor to British gardens, however it's not seen in the northern and western isles of Scotland.

Identification
The mistle thrush has round black spots on its breast while its belly is much paler than the rest of the body. The upper parts are a grey-brown and the under parts of the wings are white.

Song
It utters a loud distinctive alarm call when its nest is under attack, much like a football rattle. The song is loud and far-reaching, and it has long pauses and brief phrases.

Nesting
The female mistle thrush builds the nest, normally in the fork of a tree or in shrubs, but sometimes in walls. The nest is made of many different substances, including earth, leaves, moss and grass. It's a neat cup-shaped construction, though rather bulky.

Length
27cm (11in)

Did you know?

■ The mistle thrush sings earlier in the year than most other birds.

■ The mistle thrush is sometimes known as a missel thrush or stormcock, after its distinctive call in the night or during bad weather. It'll happily sing in all weathers, even hard rain.

Moorhen
Gallinula chloropus

Moorhens are hardly regular garden birds, but if you live near a pond or canal with resident moorhens, they may make the occasional pilgrimage to your place to check out any easy peckings on offer. The moorhen is amphibious, feeding in water and on the surrounding grassland. It eats a mixed diet of leaves, seeds, berries, worms, snails, tadpoles and tiddlers, and other birds' eggs. In the garden, a moorhen takes bread and seeds that have fallen on to the ground from a bird feeder.

The enormously long toes that spread the moorhen's weight when it's walking over mud or on water, from lily pad to lily pad, look very awkward on dry land. The moorhen steps very deliberately, almost in slow motion, like a diver in flippers plodding about out of water. If it looks out of sorts on land, it seems even more incongruous when standing on top of a hedge or in a tree. However, appearances can be deceiving: the moorhen is, in fact, hugely surefooted as it climbs and clambers about in the branches.

A visit to your local pond will give the moorhens a chance to show you what they do best – swim smoothly across the water, head jerking forwards and backwards in time with a flicking tail. You may witness the odd splashy skirmish between two moorhens competing

A moorhen is generally reluctant to take to the air; if panicked into flying, it runs across the ground or water, fluttering its wings until it's airborne, then flaps feebly away with its legs hanging down, and touches down again as soon as it can.

Facts

Frequency
Most quiet stretches of still or slow-moving water with reed-lined banks in lowland Britain are home to moorhens at some point in the year; approximately 270,000 breeding pairs.

Identification
Both male and female are bluish-black with white feathers under the tail and along the flanks; the adult has a bright red facial plate and beak with a yellow tip; long greeny-yellow legs with scarlet garters at the top and huge long-toed feet at the bottom. It hatches as a sooty-black chick with a bald red head and red beak, but the colour fades until the juvenile is pale greyish greeny-brown.

Song
Frequently uses coarse, noisy *kittik* and *kurruk* alarm calls.

Nesting
The male weaves several rafts from reeds and other water plants in a secluded location near water; once the female chooses her favourite site, the pair works together to weave a secure island with a deep cup in the centre to hold the eggs and chicks.

Length
32-35cm (13-14in)

'The moorhen, too, as proud as they
With jerking neck is making way.'
From 'The Pasture Pond' by Edmund Blunden

or their own space. For a bird that seems to be so afraid of its own shadow, the moorhen can be pretty hot-headed, especially during the breeding season; rival birds strike out violently at each other with their feet and even try to drown one another.

In spring and summer, you may be able to spot a raft-like nest of leaves and stems tucked away among the bankside vegetation. With any luck, you may see a family of newly-hatched chicks later on as well – the cutest sight on a summer pond. Within hours of hatching, the tiny balls of black fluff are swimming about the pond like downy dark bubbles.

Did you know?

■ Moorhens employ family helpers to allow them to raise up to three large broods a year. Older offspring from the first brood stay around to assist their parents in finding food for their younger siblings. Supporting three or four generations at the same time can severely test the quality of the original territory.

■ The moorhen has a great vanishing trick. It rarely dives for food, but when danger threatens it doesn't just hide in the bankside vegetation; it disappears under water, hanging on to weeds and waterside vegetation with its feet to keep itself submerged. Only its beak breaks the surface to let it breathe while under water.

Nuthatch
Sitta europaea

If you see a bird climbing the trunk of a tree, there are a number of options. It could be a woodpecker, or perhaps a treecreeper. If you see it coming back down again, then you can narrow the list down to one. It's a nuthatch.

Remarkably, this woodland bird, that can often be rather difficult to see and tends to keep itself to itself, is a surprisingly common garden visitor. Peanuts are your best bet for enticing it in. It will fly down to the feeder, prise out a nut, then fly off with it to consume at its leisure. In addition, if your garden is often frequented by large flocks of blue and great tits, it's worth scanning the foraging party for the occasional nuthatch that's come along for the ride.

And then there's the bird's name itself. If you have hazelnuts in your garden, and a visiting nuthatch, watch the way that the bird jams the nut into a slight crevice in a branch, then, once it's wedged in place, hammers away at it, using its entire body weight. Once the nut has been split open, the nuthatch feasts on the edible kernel inside. The banging is highly audible, and is another good clue that a nuthatch is living somewhere near your garden, even if you can't see it.

Facts

Frequency
Surprisingly common in gardens up and down the country.

Identification
Slate-grey upper parts and a peachy belly are the best clues to identifying this stocky little bird. Look, too, for its black eyestripe and, in flight, blunt tail with white notches on the outside.

Song
A ringing *pee-pee-pee-pee* song, along with a number of high-pitched *seeps* and *kiks*.

Nesting
A tree-nester, the nuthatch will narrow the opening of the tree hole by plastering it with mud, so that any birds that are larger than itself can't get in.

Length
14cm (5½in)

Although it looks like a woodpecker, the nuthatch is, in fact, a member of a completely different family.

Did you know?

■ The nuthatch lines its nest so thoroughly with mud that, once it's set, it's extremely difficult to prise away.

■ Don't go to Ireland in search of nuthatches – there aren't any there.

Mike Russell, Sussex Wildlife Trust, advises:

■ One bird that can really add a splash of colour to your garden on a cold wintry day is a nuthatch. This small woodpecker-shaped bird with its lovely grey-blue back and chestnut front explodes onto the bird feeders, grabs an item of food and then retreats as quickly as it arrived to eat it somewhere safe. It'll often wedge the food in a fork in the tree and then use its dagger-like bill to cope with its major food items, such as hazelnuts, acorns and seeds, but it's also partial to the occasional insect.

■ The good news is that nuthatches are now doing well and the population is expanding northwards, so the chances of them appearing in your garden, up as far as southern Scotland, is increasing. It's essentially a woodland bird but has adapted readily to parks and gardens that contain large and mature trees. Feeding has also encouraged the nuthatch into gardens and now, for many, it's a frequent visitor.

■ The nuthatch likes to nest in holes in trees, so is quite happy to use nest boxes, and you can often tell when a pair is in residence as they apply a lining of mud to the entrance hole to make it exactly the right size for them to squeeze into. During the early spring its distinctive call echoes around the woodlands which continues until it has young to feed, by which time it becomes silent and can be quite difficult to see.

■ Unlike most other birds, the nuthatch is able to move down the tree trunk, which is very characteristic of its behaviour and is a good way of telling it from other garden or woodland birds. But a good view in a good light makes this lovely resident a very easy bird to identify.

> **'You intent on your task and I on the law**
> **Of your wonderful head and gymnastic claw'**
>
> From 'The Nuthatch' by Edith M Thomas

Pied Wagtail
Motacilla alba

This is a book about garden birds, but if there were a special chapter reserved for 'very urban birds', then the pied wagtail would probably take the opening pages. Pied wagtails are those familiar and confiding fellows that are often seen in places where most other birds fear to tread. Supermarket car parks, train station platforms and even the busiest of high streets all frequently play host to this black and white little insectivore, with a flicking tail that nips in and about and around people's feet, on the hunt for its next snack.

When the pied wagtail spots an insect, it runs quickly towards it with a twinkle-toed dash, leaping and twisting in the air to make its catch. They've adapted extremely well to the march of urbanisation, and have made towny insects their own.

Pied wagtails are quite at home in and around gardens, sometimes dropping down to lawns to

Facts

Frequency
There used to be over 500,000 breeding pairs of pied wagtails in the British Isles, but the bird has suffered a decline of up to 50% in the last few decades, particularly in rural areas, where herbicides have reduced its feeding opportunities.

Identification
It's pied and it wags its tail – quite simple, really. Look out, though, for white wagtails, of which the pied wagtail is currently seen as a subspecies, but the two might be split into separate species one day. The white wagtail is a paler bird, with a grey back rather than black and, although it lives on the continent, it occasionally turns up in Britain when it overshoots its migratory path.

Song
A loud *sziszit*, emitted when in flight and during courtship.

Nesting
The pied wagtail is a flexible nester, using twigs, leaves, grass, hair, feathers, wool and anything else it can find. It's pretty flexible about its nesting site, too, and has been recorded nesting in all sorts of crevices, from walls and nest boxes to old machinery.

Length
18cm (7in)

Did you know?

■ The pied wagtail is quite a territorial bird in the breeding season, and has been known to attack its own reflection in windows.

■ In exceptionally cold winters, it can't afford to be choosy about where it hunts its insects, and sometimes ends up at sewage farms where the prey is still in reasonable supply.

■ Wagtails roost in large numbers – hundreds and sometimes even thousands – quite often close to industrial sites, or even in trees in busy high streets.

continue their busy pursuit of the next morsel, sometimes hanging around rooftops, especially on sunny days, where insects settle on the tiles to enjoy the warmth of the sun. So adept have they become at hunting, that where other insect-eaters such as warblers largely migrate south in the winter, the pied wagtail stays on, searching the hibernating insects' hidey-holes. At this time of year their greatest success comes from the margins of watery areas, so if you have a pond in your garden there's a chance you'll see a wagtail flicking around its edge. In very cold weather, they'll even come to bird tables in search of mealworms or chopped peanuts.

You'll also hear pied wagtails come in. They almost always emit a loud *sziszit* call as they fly, and when you look up you'll see this confident little bird swooping up and down, its long tail stuck out firmly behind it, ready for a quick wag once it lands on your lawn.

'He stooped to get a worm
And looked up to get a fly.'
From 'Little Trotty Wagtail' by John Clare

When you're mowing your lawn, keep an eye out behind you. There might be a pied wagtail trotting along, picking up the insects you've disturbed.

Red Kite
Milvus milvus

Reintroductions are difficult. Once a species has become extinct in a country, other animals move in to fill the niche, and trying to bring the animals back is by no means a certain success.

In the case of the red kite, however, it's been a great success. This magnificent bird, with its broad wingspan and elegantly-forked tail, was once a common bird of prey in urban Britain, and a much appreciated one, too. A scavenger by nature, it helped to clean the streets of rubbish and by the 17th century its hunting grounds in London were even protected by royal decree, so useful were they at keeping down disease. A century later, however, drainage systems and better hygiene kicked in and the kite, needing to move elsewhere for its meals, began to lose its favoured status. Taking to the countryside,

it soon became hated by gamekeepers and farmers alike, who accused it of stealing their stock and, by leaving poisoned carcasses lying around, they rapidly saw off the bird.

As the 19th century drew to a close, the red kite had been hounded out of both England and Scotland, and had just one remaining stronghold in the valleys of mid Wales. The situation was no better in the next century, but in the 1990s an ambitious new plan was hatched to bring the kite back – to introduce birds from Spain and Sweden to carefully selected sites in Scotland and the Chilterns. Within a few years they were breeding and, today, the number of successful pairs is reaching several hundred and climbing.

To see these amazing birds wheel slowly above your head is an awesome sight, and

Red kites sometimes decorate their nests with flowers, possibly serving as a disinfectant, or a cooling system.

Facts

Frequency
There are now some 500 breeding pairs of red kite in Britain, at various sites in the Chilterns, north and south Scotland, the east Midlands, Tyneside and Yorkshire, as well as the original Welsh birds themselves, of course.

Identification
The forked tail of the gliding kite is its most distinctive attribute and, along with the bird's bent 'wrist' in its wings, it will enable you immediately to know that you're not looking at a buzzard. If the bird is grounded, look for the rich chestnut plumage and pale-grey head, both streaked with black. Males and females look alike, although the latter is slightly larger.

Song
The kite sounds almost like a flying cat, emitting a slightly haunting *meow* call, which becomes cacophonous when a low-flying group is calling together.

Nesting
The red kite pair will tend to seek out an old crow's nest in the fork of tree and build a platform of twigs on top of it. They'll then line it with virtually anything they can find, including wool, fabric and even plastic bags.

Length
60-66cm (24-26in)

many homeowners are tempted to leave food out for them to entice them into their gardens. It's advisable not to, though. Although there are feeding stations at some of the birds' strongholds, these are carefully managed by experts who know what the kite should be feeding on. These newly-established birds are not as robust as their predecessors of centuries past and old food scraps, particularly cooked meat, may make them ill. Better instead to watch them wheel in the sunshine as they concentrate on building their numbers back up to sustainable levels once again.

Did you know?

■ When reintroduced red kites nested in England in 1992, it was the country's first breeding record for 122 years.

■ During the courtship season in March, red kites are a sight to see. The male and female play together by flying quickly towards each other, pulling out of collision at the last moment, sometimes even clashing talons in the process.

'The kite, tiw-whiw, full oft Cried, soaring up aloft.'
From 'Evening Walk' by William Wordsworth

Redwing
Turdus iliacus

If ever there was a reason to leave fallen apples on the ground in late autumn, then it's the redwing that provides it. This small thrush, which leaves its Scandinavian home to escape the harsh winter months there, arrives exhausted on our shores in October and needs quick re-fuelling. Fallen apples, as well as full berry trees and shrubs, can provide it with just the pep it needs to pick itself up. Then by the spring, it's off back to its homeland.

A very wintry symbol, the redwing (so named for the flash of deep orange under its wings) occasionally stays over into the summer in Scotland, where it has even been recorded as breeding. Among the birch woodland and conifer plantations of the Cairngorms, and even in Orkney and Shetland, a few dozen pairs stick around each year and bring up their young.

You might see them in your garden, even if you don't have fruit or berry-bearing plants. Redwings arrive in such great numbers, that they may stop off wherever generous gardeners leave fruit out for them, particularly in the colder winters. In Kent, nighttime fishermen have reported eerie whooshing winds while at work – the sound of thousands of redwings migrating during the night, their flight calls sounding like wind rushing through the trees.

Facts

Frequency
As many as one million redwings migrate from Scandinavia to Britain each year.

Identification
It's the smallest British thrush, with a dark chocolate back and wings, heavily-streaked breast and distinctive white eyestripe. The bird's creamy underparts reveal rust-red flanks in flight.

Song
The redwing emits a *whseept* call when in flight, and a soft song in early spring before the trip back home.

Nesting
Twigs are glued together with mud in a tree fork or shrub, then it's lined with grass and lichen. On occasion, the redwing will nest on the ground.

Length
21cm (8in)

Did you know?

■ Hardy souls, redwings have been recorded as living for as long as 18 years.

  Mike Russell, Sussex Wildlife Trust, advises:

■ A bird appears in your garden during the winter, which looks like the resident song thrush, but there is something different about it. It's slightly smaller and has a distinctive stripe just above the eye and, as it flies off, a flash of orange-red is revealed on the flank which clinches its identity – it's a redwing.

■ The redwing is a common winter visitor to the UK – the majority arrive in October – escaping from the harshness of its breeding grounds in Northern Europe. It often arrives in the hours of darkness and the sharp-eared among you might pick up its call as it passes over your head. For over 50 years now a few pairs (fewer than 100) have remained to breed, principally in central Scotland.

■ When it arrives in Britain it prefers open countryside, farmland, orchards – anywhere there are hedges full of berries and fruit. It will often associate with its bigger Nordic cousin, the fieldfare, and flocks of several hundred mixed thrushes aren't unusual throughout the country.

■ As the supply diminishes and winter tightens its grip, the redwing has to search further afield to find food which is when it's likely to turn up in your garden. If you have any berry-bearing bushes, such as holly, *Cotoneaster* or *Berberis*, these will prove very attractive to redwings; large flocks may descend and strip your bush in a very short space of time. You're more likely to encounter a redwing after the turn of the year and its visit may be short-lived as it will move on once it's eaten all your berries.

■ Come March, the redwing will be thinking of returning to its northern breeding grounds and your garden will be left to the more-familiar song thrush.

Although redwings defer to mistle thrushes at feeding stations, no thrush is a match for a full flock of hungry redwings.

Cats

Royal Horticultural Society

Cats are not just the foe of birds – they can destroy a gardener's hard work, too. Here, Fiona Dennis of the RHS provides some suggestions for how to keep unwanted moggies out of your garden.

Much-loved pets can be the bane of nearby gardeners, whose plots they use as toilet areas. Holes are scraped in flower and vegetable beds, particularly where the soil is newly cultivated, and vulnerable young seedlings can be destroyed. Gravel paths prove attractive sites for defecation. The local tomcats will also scent-mark their territories by spraying urine on plants, which can scorch foliage. Damage to the bark of trees and shrubs, caused by cats sharpening their claws, is another form of territorial marking. Cats also have a habit of sunbathing on their favourite plant, *Nepeta faassenii*, or ornamental catnip, crushing plants in the process. Problems are most severe in high-density housing areas, where cats are numerous and gardens small.

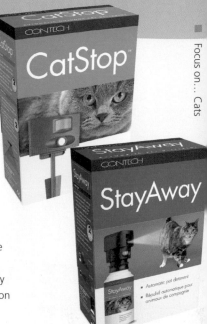

Repellents

These include products containing pepper powder (Bio Pepper Dust, Secto Pepper Dust) and essential oils (Growing Success Cat Repellent). Such repellents give only short-term protection and need frequent reapplication. Remove any cat excrement before use. A physical repellent is Catscat, which is a mat made from a web of soft bendy spikes. A cat repellent plant, sold under the names of Scardy Cat® or *Coleus canina*, is available from Thompson & Morgan, Poplar Lane, Ipswich IP8 3BU, tel: 01473 688821. The foliage produces an unpleasant smell when touched. This plant can be grown outdoors in the summer, but needs frost protection in the winter and over-wintering in a cool greenhouse. Another kind of repellent, which is motion activated, is a water jet (such as Scarecrow, sold by Rockwell products). In both instances, they need to be moved around in order to catch the cats, who will otherwise learn a 'safe' route past the repellent.

Electronic devices

These are mainly sold by mail order, so look for advertisements in the RHS magazine *The Garden* and other gardening magazines. Most produce ultrasonic sound (barely audible to human ears) when triggered by a motion sensor. Some cats flee when they come within range, while others, perhaps the more dominant local cats, hold their ground and carry on regardless. The best results are in open gardens where the ultrasound is not baffled by shrubs or fences. Place the speaker at one end of the garden, as sound travels away from the device in the direction it is facing.

Deterring cats

Cats roam freely through their territories and are too agile to be excluded by fencing or netting. However:
- netting may help keep cats away from small areas within the garden
- flower borders densely planted with perennials are less appealing as toilet areas – no bare soil to scratch
- keep seed rows well watered, as cats dislike wet soil, preferring loose, dry earth and mulch
- use one or more of the cat deterrents on the market. They fall into two groups: repellents that offend the cat's sense of smell or taste, such as cocoa shell mulch, and electronic scaring devices. Neither type causes harm.

Cats are responsible for many millions of bird deaths every year in Britain's gardens. If you have a cat, you can help protect birds by making sure it wears a collar with a bell, and by feeding it when birds are at their most active in the garden – around dawn and dusk.

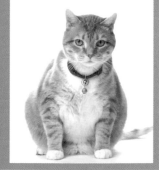

Reed Bunting
Emberiza schoeniclus

During the winter, if you happen to spot a natty, slimline, sparrow-looking bird that looks slightly out of place in the flock of sparrows eating seeds from the ground in your garden, the striking visitor is worth a second look. It could be a reed bunting, in his subdued brown winter colours. As its name suggests, for most of the year the reed bunting's life is focused on wetlands and waterways, where it feeds among the reedbeds, rushes, sedges and willows that grow along the banks of rivers, streams, canals and dykes and around ponds.

In the autumn, reed buntings roam more freely, away from their watery haunts, seeking out seeds and grains on farmland as well as marshland. Sometimes, small bands of them link up with flocks of yellowhammers and finches, which are also combing the countryside for seeds. For small birds, there's safety in numbers, when they can roost together, huddling up close to keep warm, especially on frosty nights. Occasionally, their wanderings bring them into gardens where there is a plentiful supply of the seeds and grains they crave. Even hungry reed buntings rarely land on bird feeders but they do take seed that's been scattered on the ground, under hedges or on patios.

An increase in garden bird-feeding is quite possibly the best news reed buntings have had in ages. Elsewhere, they've been having a hard time over the past 30 years or so. They've seen their natural reedbed habitat greatly diminished by changes in land drainage and as a result of neglect – once, reedbeds were cut regularly to provide reeds for thatching roofs, which encouraged vigorous new growth every year, keeping other marsh-loving plants at bay, but with the fall in demand for thatch, boggy woodland species started to supplant the reeds. In addition, fewer wetlands and a heavy-handed use of pesticides and fertilisers meant there were fewer caterpillars, flies, beetles, aphids, bugs and spiders around to feed themselves and their chicks each summer. To top it all, a shift to sowing cereal crops in the autumn, rather than leaving the stubble fields fallow until the following spring, left the reed bunting with less grain to keep it going over the winter.

Did you know?

■ Recently, reed buntings have been found nesting in surprisingly dry places, such as young conifer plantations and oilseed rape fields. In some areas, up to 75% of reed bunting families are now reared in fields of rape. Unfortunately, they may still have second broods in the nest when it's time to harvest the rape. Before harvesting, the plants are killed and left to dry for a few weeks. This can be done either mechanically with a flailing machine, which inevitably smashes the nest, or with a herbicide spray, which gives the reed bunting chicks a fighting chance of fledging before the crop is cut. For once, agro-chemicals are not such a bad thing.

■ The reed bunting's eggs are beautiful; the pale glossy shell is tinged olive, brown or purple and covered in Jackson Pollock-style purple-black streaks, spots and blotches.

■ Your best chance of seeing a male reed bunting at his dandiest is to take a walk beside a reed-lined waterway in spring. Then it's hard to miss his smart black head when he perches high up on a reed stem or willow tree, twitching his tail and wings, to recite his tinkling ditty time and time again.

■ The reed bunting is on the Red List, as there was a decline in population of over 60% between the 1970s and 2000. The good news is that major reedbeds are now managed by conservation agencies. New reedbeds are being planted to purify sewage, industrial effluent and contaminated drainage from farmland. Underwater, the reeds trap and oxygenate the sediment, while bacteria in the silt break down the waste and clean up the water. It's hoped that these new reedbeds will eventually become home to a number of rare wetland birds and more reed buntings. Farmers are also being sponsored to leave wider field margins and plant seed-bearing crops on set-aside land, to supply dispossessed birds such as the reed bunting with places to nest and plenty of food over the winter.

Facts

Frequency
Still fairly common in its usual waterside haunts throughout most of the British Isles; the current estimate of breeding population is 210,000 pairs. Nests are built low down and often trampled on, which means many eggs are failing to hatch, chicks are taken from the nest by predators and there's a poor survival rate among young reed buntings, maybe because they can't find enough to eat as inexperienced youngsters or during their first winter.

Identification
The male in summer has a striking jet-black head, throat and bib with a white collar and drooping moustache-like stripe on his beak; his upper parts are rich reddish brown with darker brown streaks; his under parts are pale with thin brown stripes. In winter, his black areas are browner; the female is brown with dark cheeks, pale 'eyebrows' and a double pale-and-black moustache and dark streaks on her breast. The juvenile looks like a streakier, paler female. They all have a long, dark, deeply-notched tail with broad white edges, conspicuous in flight.

Song
The male is a chatterbox, singing *srip-srip-srip sea-sea-sea stitip-itip-itipip* many times during courtship and breeding.

Nesting
The female weaves an untidy cup of dry reeds and grasses which she lines with hair, finer grass and old reed tops and willow fluff, low down near water, among reed stems, in an old willow stump, grassy tussock or clump of nettles.

Length
15-16cm (6in)

Ring-necked Parakeet
Psittacula krameri

Occasionally, when identifying birds, the word 'unmistakable' is used. Rarely, however, is it used with as much confidence as in the case of Britain's one and only parrot. If you live in parts of south-east England, and probably in the years to come somewhat further afield, then the ring-necked parakeet will not fail to register itself in your consciousness.

Bright green, squawky and manic, the parakeet was once just one of many types of parrot that filled domestic aviaries across the country. Pet tropical birds often escape, but are rarely able to establish themselves in Britain, thanks to the weather, but the ring-necked parakeet gained a small foothold in the 1970s which, with the milder winters of recent years, eventually became full-blown colonisation.

As yet, numbers are small enough that their presence provides less of a threat than a treat to gardeners, as they noisily dash in and out of gardens enjoying the fruits, seeds and nuts on offer. They've developed an omnivorous diet in this country, though, and are sometimes seen picking over household scraps and even meat, suggesting that they could expand quite rapidly in the years ahead.

This, of course, means that they've got a very good chance of being seen as a pest somewhere down the line. Rapid colonisation invariably affects native creatures, and the woodpeckers and owls, whose holes the parakeets also seek for nesting purposes, may one day suffer.

For now, though, ring-necked parakeets are eye-catching garden visitors. Love them or hate them, you can't ignore them.

Facts

Frequency
Now numbering several thousand individuals in the south east, the bird's numbers are likely to grow rapidly if winters continue to be mild.

Identification
Bright green, long-tailed and with a red bill, the parakeet looks like no other British bird. It flies in a straight, direct line, screeching as it goes.

Song
A parrot's screech, along with a number of chattering calls.

Nesting
Tree holes are its nesting sites, which it tends to occupy more than most native British birds.

Length
40cm (16in)

Did you know?

■ The ring-necked parakeet's population is largely centred around London and the south east, but sightings have been recorded in almost every county in England, and have reached both the Scottish and Welsh borders.

■ The bird (also known as the rose-ringed parakeet) has established feral populations in many parts of the world, including California, Florida, and parts of South Africa, Germany, France and Japan. Not bad for a bird that originally only called South Asia home.

Here's the proof that, in one view at least, Britain's gardeners are pigs. Centuries ago, when wild boar freely roamed the nation's countryside, its frequent companion was the robin. As the pig rooted around in the undergrowth, turning over the soil in its hunt for food, so the robin hopped alongside, picking out the choice insects and worms that emerged. Eventually, of course, the wild boar disappeared from these shores, but the robin didn't mind. It soon found a replacement for its porcine provider – the spade-wielding gardener.

This may not perhaps be the most poetic of reasons for the apparent friendliness of the gardener's little partner with the bright red breast, but not to worry. Robins are Britain's best-loved birds, and their aggressive nature, territorial tendencies and ever-rampant appetite will never detract from their status. In fact, it may even enhance it. Their images appearing on everything from Christmas cards to pencil tops, robins have been embraced by a nation that loves them for their boldness, simplicity, companionship and year-round presence. They're as traditional as the yeomen of old England themselves.

To cap it all off, the robin's song is perhaps the most beautiful of them all. The nightingale may be the famed songster, but next time you hear a robin strike up in your garden, sit back, close your eyes and listen. Wistful and melancholic during the winter, yet confident and powerful in the spring, it's a song that perfectly reflects the changing of the seasons.

Not all robins will prance around your feet, however. During the winter months, those that arrive with the continental influx tend to be more skulking and shy, flitting among the foliage like an out-of-season warbler. Further hiccups are sometimes made with the young, who exhibit not a hint of red, but a darkly speckled breast instead.

Facts

Frequency
Resident all year round across the country. One of our commonest birds, its numbers probably top four million territories.

Identification
It's almost unnecessary to reveal that the robin can best be identified by its bright red breast. It is worth mentioning, though, that males and females look the same, while the young sport speckled breasts.

Song
Delivered all year round, although with a sadder tone during the winter months. Quite often heard at night, particularly if the bird is roosting near a street light.

Nesting
The robin is famous for nesting in virtually anything with easy access. Boots, old garden-coat pockets, even car bonnets and watering cans have all at one time or another carried the house name 'Dun Robin'. The nest itself is made of leaves and moss and lined with hair.

Length
14cm (5.5in)

Did you know?

■ Both male and female robins sing.

■ In a good year, with an early spring and long summer, robins can rear up to four broods.

■ In some parts of the country, up to 10% of robin deaths are caused by territorial fights.

'A robin redbreast in a cage
Puts all heaven in a rage'
From 'Auguries of Innocence' by William Blake

Mike Russell,
Sussex Wildlife Trust,
advises:

◼ When it comes to nominating Britain's favourite bird, the robin usually wins hands down. This little bird with the orange-red breast somehow captures the essence of our native wildlife, particularly during the dark days of winter when its cheery song can brighten up many a gloomy day – perhaps that's why we love the robin so much. They are obvious, inquisitive and often suddenly appear as soon as you go into your garden.

◼ But robins do have a bit of a dark side as many people have witnessed. Unlike other small birds they maintain a winter feeding territory and if your garden is a good source of food then the robin is going to defend it vigorously, very occasionally even to the death. That's why you can hear robins singing throughout the winter as they proclaim their feeding territory. In fact the only time they are quiet is just after the breeding season when they go into moult and are at their most vulnerable to predators.

◼ Another difference that marks robins out from other garden birds is that during the winter months the female sings as well because she too needs to hold onto to her valuable food source.

◼ Your garden can provide important nesting sites as well. Robins usually need shelter and an ivy covered fence or thick shrubs can provide the ideal location. They will readily use open-fronted nest boxes in cover and have been known to requisition unusual objects, such as old kettles, in which to raise their young. To see the speckled young robins chasing their hard-working parents around the garden is a lovely sight, and a great reward for those who try and provide a helping hand for Britain's favourite bird.

Robins have such strong parental instincts, that they have been recorded feeding the nestlings of other species.

Rook
Corvus frugilegus

The rook is a highly sociable bird that likes the company of its fellows, but doesn't seem so keen on the company of humans. Perhaps that's unsurprising considering the number of rook pies that were consumed until recent years. Rarely coming into towns, they build their large colonies, or rookeries, near farms and farmland, sometimes in huge communities of up to 1,000 nests, built at the tops of tall trees. These rookeries can last for years: if you remember one from your childhood, go back and have a look – if the trees are still there, the likelihood is that the rooks' nests are too.

The only real chance you have of seeing a rook in your garden is if you live near to a rookery in a village or small town. But if you do, as they're so noisy you'll hardly need to have it pointed out to you. One of the reasons is that rooks are very attentive birds to their homes. While other species abandon their nests during the wintery months, finding better things to do, rooks often stay in the vicinity, even tickling up the odd stick here and there as they make running repairs. That's good nest management.

When not in the breeding season they congregate in large colonies, often in fields, during the day to feed on a varied diet including invertebrates, small mammals, grain and vegetable matter. Then they meet up at dusk with other large colonies of other rooks, jackdaws, crows and even sometimes magpies, to roost in the same place every night.

Yet large towns hold nothing of interest to them. It seems odd that a bird that seems quite at home by the side of the motorway, picking away at a bit of old roadkill and oblivious to the traffic hurtling past them at fatal speeds, should be so wary of conurbations, but then their omnivorous diet has never forced them to look

Although rooks are very sociable birds, they frequently steal twigs from each others' nests to make their own.

Facts

Frequency
Some 950,000 breeding pairs can be found across all of Britain and Ireland, except north west Scotland.

Identification
The plumage is all black with a reddish or purplish gloss but around the base of its beak – nostrils and chin – is bare skin which can look whiteish. The untidy appearance arises from the slightly peaked head and the thigh feathers, which look like baggy trousers. The bill and legs are black. Juveniles do not have the bare skin at the base of the beak so can be mistaken for carrion crows.

Song
Raucous *caw, caws*.

Nesting
Rooks nest in a colony called a rookery. The bulky nests are built high in the treetops and are made from twigs bound together with earth, lined with moss, leaves, grass, wool, hair etc. Previous years' nests may be renovated and reused.

Length
18cm (7in)

further afield than the countryside. Besides, if you spend your time in towns there's not going to be room for all your friends, and rooks do like to be with their friends.

One drawback about being just one in a crowd is that you have to find a way of standing out from the rest. Once the breeding period arrives, male rooks spend a lot of time strutting around in front of their prospective partners, bowing, posturing and cawing and, if he senses success, emptying the contents of his food pouch into her mouth before mating can take place. Very touching.

'Sooth'd by the genial warmth, the sawing rook
Anticipates the spring, selects her mate,
Haunts her tall nest trees, and with sedulous care
Repairs her wicker eyrie, tempest torn'
From *The Natural History of Selbourne* by Rev Gilbert White

Did you know?

■ Like many members of the crow family, the rook figures heavily in folklore, and the sudden desertion of a rookery was said to be a bad omen for the landowner. Rooks are believed to indicate rain through their behaviour, and legend has it that they are also able to smell approaching death. It's amazing how a black plumage can create such gloomy folklore.

■ Young rooks will stay with their parents for about five months then set off by themselves. It is not unusual to see large groups of young birds swooping together in the autumn sky.

...ius

Unless you have pine trees in your vicinity, you're unlikely to see these little finches during the summer months. But when winter comes around it's a different story. Siskins have spread their range dramatically in the last half-century, moving beyond the conifer plantations of their once Scottish-only breeding grounds, and nesting as far afield as East Anglia, Devon and mid-Wales. In addition to such rapid growth, a further half a million birds from Europe spend their winters here, so that up and down the country, over a million of the birds are in search of food.

Enter the humble garden. These birds that once kept themselves tucked away in copses and thickets during the winter months have become the latest additions to bird-table entertainment. Smaller, slightly yellower, and with more dark streaking than greenfinches, they have adapted their abilities for hacking into pine cones to enable them to take what they need from peanut feeders. They may not be as big as some of their competitors, but they've got courage to make up for their size, and have no qualms about seeing off other birds once they've installed themselves.

Such competitiveness is born of necessity. During the summer months when food is plentiful, siskins largely keep themselves to

'**And the small mad siskins flit by,
Flying upward in little skips and
erratic leaps;
Or they sit sideways on limber
dandelion stems,
Bending them down to the ground.'**
From 'The Siskins' by Theodore Roethke

Facts

Frequency
The siskin can be found all year round in northern England, Wales, Scotland and Ireland. In southern England, it's mainly only seen as a winter visitor from the continent.

Identification
Broad yellow bars across the wings are shared by both male and female, but the former also boasts a black cap and bib and bright yellow breast during the summer months. The female is brown and streaky.

Song
The song is high trill which drops at the end to a low buzz. The siskin also employs a number of chattering calls when in flocks.

Nesting
A complex structure of twigs, grass, moss, feathers and hair, the nest is usually placed on the outermost twigs of a pine tree branch, for safety.

Length
12cm (4.5in)

It has been noticed that siskins are more attracted to peanuts in red nylon nets than in anything else. This is perhaps because the bags remind them more of the shape and colour of pine cones.

themselves in their wooded areas, finding their niche and settling into it. Come winter, though, it's every bird for itself, and siskins have quickly learnt how to play that game.

You may also find siskins in your garden if you have catkins. Small and agile enough to be able to reach into the catkins and pull out its seeds, the siskin twitters its way through the alders while other birds can only look on.

The bird's song has given it many problems in the past, though. Bright and yellow, it was frequently caught in the 19th century, as it flew from Europe, and sold as a cheap alternative the expensive and more prestigious canary.

Did you know?

■ Siskins often mix with redpoll flocks as they search for food.

■ The female's golden wingbars are very slightly narrower than her mate's.

183

Do you have an anvil in your garden? There's a good chance that you do, although in this case it has nothing to do with blacksmiths. 'Anvil' is the name given to a song thrush's favourite stone, which it smashes snails against to get at the mollusc's flesh. If you do have an anvil, you'll know by the litter of broken shells around it. Little wonder then that the snail-devouring song thrush is one of the gardener's best friends. Although earthworms form the bulk of the bird's diet, in colder weather when the ground is too hard to get at them, snails and slugs will do just as well instead. For this reason, it's best to avoid using slug pellets in your garden, and let the song thrush take over as your very own natural pesticide.

The song thrush is well named, striking up its tuneful courtship melody as early as January, having fought for and prepared its territory at the end of the previous year. Like its fellow thrush, the blackbird, it has a wide repertoire of short phrases, but you can tell the two birds apart because the song thrush tends to repeat each phrase.

Not as brassy as the mistle thrush with which it is often confused, the song thrush is a more earthbound bird, frequently heard but not seen flipping leaves about under bushes and shrubbery as it searches for invertebrates to eat.

> 'That's the wise thrush; he sings each song twice over
> Lest you should thing he never could recapture
> The first fine careless rapture.'
> From 'Home Thoughts from Abroad' by Robert Browning

Frequency

The last three decades of the 20th century saw song thrush numbers collapse, and today there are about one million breeding pairs across Britain (except for remote parts of northern Scotland). This still makes it one of Britain's most common species, but the population is well short of the several millions of birds that once lived here. Each year, some birds fly south to winter quarters around the Mediterranean.

Identification

Nut-brown upperparts help to distinguish the song thrush from its close relative the mistle thrush, as does the more cleanly-speckled white breast, and the flash of orange under the wing when the bird takes flight. The song thrush is also slightly smaller.

Song

A series of beautifully crafted fluting phrases, mostly sung twice. Like a blackbird, the song thrush emits a sharp *chick chick* alarm call. In flight, a softer *tsip* can be heard.

Nesting

The song thrush nest, constructed several metres above ground in a tree or bush, is well built and sturdy. Bowl-shaped, it is first formed from grass, twigs and moss, and then plastered with mud, dung or saliva for strength.

Length

23cm (9in)

Earthworms comprise the majority of a young thrush's diet. The hungry little nestlings can eat over 3m (10ft) of them each day!

Did you know?

■ In many parts of the country, the thrush was known as the throstle. The name is still used by those with long memories.

■ Sadly, thrushes are considered a delicacy in France and Italy, where thousands are killed each year for the table.

■ The thrush's cryptic markings enable it to remain hidden from the casual glance as it squats on its nest.

et berries are on the menu, too, particularly when colder weather keeps its main diet at bay. Rowan, hawthorn, ivy, holly and cotoneaster are all worth growing in your garden if you'd like to bring this beautiful songster in. It doesn't tend to visit bird tables very often, although if you scatter food beneath them you'll help the birds out in the coldest of winters.

The trouble with being a frequent ground feeder, though, is that you're susceptible to cats. Song thrush numbers, particularly after the breeding season, are often struck quite heavily by domestic cats that aren't even looking for food but are just testing out their hunting abilities. If you have an active cat, then you're unlikely to get song thrushes dropping in. It might be time to invest in that bell.

Keith Martin, London Wildlife Trust, advises:

The loud and proclaiming song of the song thrush an be heard throughout the day, but most regularly efore dawn and after sunset. The clearly uttered vely phrases and repetitions make the song thrush ne of the most beautiful and charismatic of our ative songbirds.

Song thrushes can potentially be found in any abitat where there is a mixture of woodland, ushes and hedgerows, a preference that often rings this species into gardens. Song thrushes est low down in any suitable cover, but typically dense shrubs, among creepers on walls or on ne ground in thick vegetation. Song thrushes feed rimarily on worms, slugs, snails and fruit.

The national song thrush population has halved nce the 1970s, although there has been a slight covery over the last decade. While this decline has been most severe in farmland areas, urban song thrush numbers have also suffered. The precise causes are not known, but are likely to be related to food supply and habitat. There are two things you can do that will go a long way to helping song thrushes in your garden:

■ Firstly, avoid using slug pellets or other molluscicides in your garden. Song thrushes prefer to eat earthworms, but turn to slugs and snails when earthworms are hard to find, particularly during spells of hot, dry weather. Slug pellets not only reduce the number of available slugs, but are also known to be toxic to song thrushes. If you insist on slug control then please use non-toxic methods.

■ The second thing you can do is to provide the type of dense shrubby habitat in your garden that song thrushes prefer for nesting. A thick hedgerow is ideal, but any planting of substantial bushes will help. This type of habitat is in decline in public open spaces, which makes the provision of nest sites within gardens an important contribution.

It's usually an alarm call and a bluster of feathers that alerts garden-watchers to the presence of a sparrowhawk, *writes Tim McGrath, Avon Wildlife Trust.* Often an observer is rewarded with a glancing view of a fast-moving brown shape hugging the contours of the garden, rapidly passing the bird table, only to disappear over a roof or hedge in a gravity-defying display of ease. Usually the lasting memory is of a staring yellow eye or the absolute silence as his magnificent aerial predator passes by. It is often noted though that after a few moments everything returns to normal and anyone missing the moment may be none the wiser.

The sparrowhawk is the top avian predator that regularly visits the garden. They are a sign of good garden health, indicating that there is enough food available for them to survive. They are not the indiscriminate killers that many headlines try and tell us, but natural predators that have evolved over thousands of years. Their predatory habits ensure that only the fittest songbirds survive. They do not hoard their food but only feed on what is needed, often their meals being days apart.

During crisp, sunny mornings in February and March, another side of the sparrowhawk can be observed, for it is during this period when territorial displays are at their peak. Orbiting above the garden both the male and the larger female will complete a heart-stopping rollercoaster flight, fluffing up their downy under-tail feathers, often calling in a high-pitched squeak that secures the pair bond.

So if you are an avid fan of feeding songbirds, you'll have to accept that sparrowhawks will also benefit from your endeavours. But rather than a bad sign, this fabulous bird is a sure indication that you're doing the right thing – so well done (and keep putting out the food)!

Facts

Frequency
There are at least 50,000 breeding pairs in the British Isles, covering the entire country except for parts of northern Scotland. Some young birds swell these ranks in the autumn when they arrive from Scandinavia.

Identification
The female is a good bit bigger than the male, and the birds' plumages differ as well. Where the male has a steel-grey back and rust-red barring under the wings, the female has a browner back with grey barring below. Juveniles are browner still.

Song
A snapping *kek kek kek* call is used for alarm, while breeding birds communicate with each other using a thin whistling call.

Nesting
First choice of tree for nesting purposes is usually a conifer. The sparrowhawk's nest is an unwieldy construct of sticks lined with tree bark, and is built in the higher branches.

Length
Males 29cm (11.5in), females 36cm (14in)

> 'The hawk slipped out of the pine
> And rose in the sunlit air:
> Steady and still he poised;
> His shadow slept on the grass.'
>
> From 'The Hawk' by AC Benson

Did you know?

■ The middle toes on a sparrowhawk's foot are unusually long. This enables the hawk to hold on more easily to small birds as it plucks them from trees.

■ Adult sparrowhawks kill about 30kg of prey per year.

■ In the early part of the 20th century, sparrowhawk populations fell into decline when gamekeepers targeted them as pests. Once they received protection, their numbers started to climb once more, only to suffer again in the 1960s thanks to DDT which got into their systems, killing some and weakening the eggs of others so that they failed to hatch.

Spotted Flycatcher
Muscicapa striata

These charming long-distant migrants from southern Africa are only on our shores between May and August, but if you're lucky enough to have one visit your garden, they're among the most fascinating birds to watch.

They're not called 'flycatchers' for nothing, and rather like a kingfisher that picks a favourite spot from which to hunt fish, the spotted flycatcher chooses a perch from which it darts into the air, twisting, turning and hovering until it has plucked an insect out of nowhere, before returning to its perch once more. During the breeding season (virtually the only time the bird spends in this country), the spotted flycatcher is able to keep several insects in its bill at a time for its young, giving you plenty of time to watch its aerial acrobatics.

Although insects form the main part of the spotted flycatcher's diet, during bad weather when insects stay low, it will either try to grab them on the ground, or even turn to berries.

So, is one likely to visit you? They are most commonly found in East Anglia where they are suited to large lawns, trees and open spaces, although their entire range covers most of Britain apart from the north-east of Scotland. Ivy provides one of its favourite nesting sites, so if you have a large lawn, a pond to encourage insects, and ivy-covered walls, you may entice in a breeding pair.

Facts

Frequency

Numbers are declining: spotted flycatchers have been given red conservation status thanks to their falling population due to changes in British woodland management, as well as problems in their southern African wintering grounds. They visit these shores from May until August, occasionally arriving as early as April.

Identification

No one could pretend that the spotted flycatcher has a distinctive plumage. It is a mousy grey-brown with a white, flecked chest and slightly streaked crown, which isn't much to go on. Its posture, however, is. An alert bird, it rests on its perch in a very upright position, unlike the many members of the warbler family with which it could easily be confused at a distance. It also tends to return to the same perch after every hunting sortie.

Song

A simple thin call-like noise and makes a thin *tsee* call.

Nesting

Both sexes make the nest and they will do so in open-fronted nest boxes. The nest is cup-shaped and made of a range of scavenged items including grass, thin twigs, lichen and spider webs, lined with feathers and hair.

Length

14cm (5.5in)

Did you know?

■ These birds do not forget a good nesting place such as a hole in the wall or a ledge and may return year after year. Sometimes the nest is even repaired in the process.

■ These birds are tolerant of noise disturbance, and use street and garden lamps as a trap to catch insects after dark.

■ The young birds roost together as a form of defence. The adults are very protective and protect nests from dive attacks from larger birds.

Spotted flycatchers are intelligent birds, and are not deterred by the warning colours on a butterfly's wings that keep other predators away.

David North, Norfolk Wildlife Trust, advises:

■ There has been a huge decline in the number of spotted flycatchers breeding in gardens and other habitats across the UK – a more than 80% fall over recent decades. The causes of this decline are not well understood, but may reflect reductions in their insect food supply, drought and other environmental changes in their African winter quarters, or most likely a combination of both.

■ These long-distance migrants arrive in gardens in late April or early May from wintering grounds in west Africa. They perch conspicuously on vantage points such as fence posts, characteristically flicking their tails before making sorties after passing insects which are caught in midair with an audible snap of the bill. The superb flying skill involved in this characteristic feeding behaviour helps identify this aptly named 'flycatcher'.

■ Spotted flycatchers will return year after year to the same favoured gardens choosing those with sheltered south-facing walls. If you are lucky enough to have flycatchers in your garden you can help them by putting up open-fronted nest boxes in sheltered sunny positions, or encouraging clematis or ivy to provide nesting cover on south-facing walls. Avoiding the use of chemical insecticides or even creating a pond will boost the insect supply which this species depends upon.

Starling
Sturnus vulgaris

Forget about CCTV covering your every move; when you get into the habit of putting out scraps for the birds, you soon learn that starlings are keeping an even closer eye on you, nearer to home. Before your back's turned, a squadron of starlings drops in to get stuck into the food. Alert lookouts and many memories to recall the best feeding stations are just two of the advantages of travelling in a flock.

But starlings are more than recyclers of leftovers; as pest controllers, they can also be a gardener's ally. To starlings, the close-cropped grass on your lawn harbours a feast of insects, earthworms and spiders. Typically, a small formation stomps over the green sward, heads down, probing the soil with their beaks. When the earth is damp and soft, a starling can part its bills underground and feel around for wriggly

Did you know?

■ If you look closely at the base of a starling's yellow beak you can tell what sex the bird is – blue for a male and pink for a female.

■ The collective noun for a group of starlings is a murmuration, which somehow grossly underestimates the hubbub of thousands of starlings gathering to roost at dusk.

■ Starlings often go back to the same nest sites year after year after year. Several pairs may nest together in loose colonies, with a number of nests in the same cavity. When starlings manage to wheedle their way into a roof space, over many generations, they can create giant nests, some of the largest of any British bird, in fact.

■ The downside to roosting in a colossal dormitory is that, in time, the droppings of thousands of starlings can damage and eventually kill the trees they rest in.

root-nibbling leatherjackets and wireworms. It can also rotate its eyes to look along its beak and gaze into a hole it's made in the earth.

Many people have mixed feelings about a mob of starlings descending on their garden; some admire their irrepressible 'go for it' approach to life and are amused by their petty wranglings, to others they're boisterous, raucous pirates, apparently bullying smaller birds out of food on the bird table. But if you think starlings make a din in your garden, then witness the sensational aerobatic display of thousands of starlings before they go to roost on autumn and winter evenings. At sunset, flock after flock gathers at an established roosting site. The sight of a vast babbling black cloud of birds whirling overhead is breathtaking. The flight is impeccably choreographed; each bird follows the others beside it so the whole crowd can swirl and swoop in perfect unison without colliding. The frenzy builds until, suddenly, the birds nosedive into cover and settle down for the night.

Facts

Frequency
Still common raiders of garden feeding stations all year round, but not numerous as they were 30 years ago, as they have suffered a decline.

Identification
Starlings appear in a bewildering variety of markings at different times of the year: after an autumn moult, males and females have white speckles and black beaks for the winter; by spring these white spangles have worn to reveal shiny black plumage that glints purple and green in the sunshine and the beaks are yellow; in early June, when hordes of strange grey-brown birds appear in the garden and start chasing adult starlings about with their beaks open, squawking loudly, demanding to be fed, it's obvious that they are the latest fledglings; as summer draws on, the youngsters' brown feathers are gradually replaced with black-and-white spotted adult winter plumage. In flight, the starling's triangular wings produce a distinctive delta-wing, 'arrowhead' silhouette.

Song
Chatter to each other in a bedlam of whistles, warbles, chuckles, gurgles, pops and clicks.

Nesting
It crams an untidy nest of straw and grass, lined with hair and feathers, into holes in trees and the walls of old buildings, or under the eaves.

Length
21cm (8in)

Pete Mella, Sheffield Wildlife Trust, and Mike Russell, Sussex Wildlife Trust, advise:

■ If you live in an area where you're fortunate enough to see the amazing, synchronised displays of vast flocks of starlings, it may be surprising to learn that these birds are in trouble. Their numbers have declined by about 66% since 1970, and these once-common birds are now on the Red Conservation List. The full reasons for this drop in numbers are not fully understood, but loss of permanent pastures, which provide insect food for their young, is thought to be a major factor.

■ The starling's natural habitat is open grasslands and it's attracted to gardens with lawns, especially those with a water supply for bathing. It has also found a niche in town centres and even rubbish dumps. Although naturally feeding on insects and their larvae, it's a very adaptable species that takes readily to garden bird tables and feeders, where it has a reputation for arriving en masse, seeing off smaller species, and devouring food quickly. This boisterousness and perceived greed has led many people who feed garden birds to discourage starlings, by protecting feeders with enclosures that prohibit access to starling-sized birds.

■ There is no doubt that starlings come across as aggressive birds and can dominate feeding stations in your garden, but if you take time out to watch and listen to them they can provide hours of entertainment. Being social birds, they spend a lot of time communicating with each other and the constant chatter of starlings gathering together in the late afternoon on rooftops and television aerials, seemingly gossiping about the day's events before setting off for a huge communal roost, is very much part of our winter soundscape.

■ Starlings nest naturally in holes and cavities in trees, but readily take to holes in buildings and nest boxes. The male often decorates the nest with leaves and flower petals. The adults, in contrast to their own liberal diet, only feed their young invertebrates, which makes a species famed for its adaptability surprisingly vulnerable if insects are in short supply.

■ In the autumn and winter months, the starling's diet changes to soft fruit and seeds, and the resident British starling's numbers are swollen by migrants from northern and central Europe, some have travelled from as far as Russia. It is during these months that they form the largest flocks and the best views can be had of their breathtaking aerial manoeuvres.

■ Starlings are expert mimics and the result of this is that you can rush out into your garden thinking you have a buzzard or yellowhammer sitting on your roof. They have also been known to imitate telephones, mobile phones and other mechanical or electrical sounds! As they are communal nesting birds, the males don't have to defend a feeding territory, but they do have to attract a mate, so if you have a wider-ranging repertoire than your neighbour you're more likely to attract a female.

■ Breeding begins in April. The process of laying the eggs and the chicks leaving takes about five weeks. When the juveniles emerge they're brown and can be observed noisily chasing their parents around the garden, demanding food until they become independent a few weeks later. Second broods are laid occasionally, so the peace you may experience in your garden is fairly short.

Swallow
Hirundo rustica

If seeing your first swallow of the summer is a delight for you, think what seeing its final destination must be like for the bird itself. This remarkable flier arrives in late March or early April, having flown all the way from southern Africa, a distance of over 10,000km. An astonishing feat for a creature that weighs no more than 20 grams. Once it's here, it immediately seeks out watery environments or other habitats where insects gather so as to replenish its weakened fuel supply. Once they've rested and pulled themselves together, they set off again to whichever part of the UK they left the previous year.

The bird's full name may be the barn swallow, but barns are by no means the sole nesting sight of this richly iridescent bird. Indeed, if you leave

> **If you water a patch of earth in your garden, you'll be providing a muddy puddle that swallows can use to build their nests.**

your garage or even shed door open in the spring months, there's a chance that a swallow will swoop in and choose it as its new home. If you have a large pond, or there's a body of water nearby, your chances are even better. Older birds will tend to return to the same nest site year after year, but each year's new crop is always on the lookout for a spanking new residence, often near where they were hatched.

Unlike swifts, swallows are perfectly capable of landing anywhere, but they still tend to remain airborne as much as possible.

Young swallows sitting on barbed wire fences or telephone wires can frequently be seen being fed by their parents, who hover in the air in front of them and pass the insect morsels across.

By the end of the summer, though, it's not just the young that are sitting on the wires. As they prepare for their long journey back to warmer climes, swallows of all ages congregate together in huge roosts, twittering frantically to each other as if to gee themselves up for the ordeal ahead. The older ones know what's ahead, but the youngsters have quite a discovery to make. The return flight can take them as long as four months, flying at about 30kph (18.5mph), and covering anything from 100km to even 300km (60-180 miles) per day. They'll have falcons and the gunfire of Pyrenean hunters to negotiate, they'll have the Sahara to cross and they'll have monsoons to weather, but many make it, and just a few months later, they're on the way back to spend their summers with us once again.

'The swallow, oft, beneath my thatch Shall twitter from her clay-built nest'
From 'A Wish' by Samuel Rogers

Facts

Frequency
Around 800,000 pairs breed in the British Isles each year, where they can be found everywhere except the Scottish Highlands and Islands. A bird of agricultural tastes, the swallow has suffered from recent changes in farm building management and use of pesticides.

Identification
An iridescent blue-black back, a ruby red throat patch and, of course, the long streamers of the deeply forked tail set the swallow apart from its fellow high fliers the martins and the swifts. A closer view also reveals ruddy underparts on the male.

Song
An enthusiastic twitter characterises the swallow's song, while in flight it emits a sharp *tswit tswit* call.

Nesting
Swallows build cup-shaped nests from mud, which they line with grass and feathers. These constructs can be found in a host of different outbuildings, from garages and barns to warehouses and even under bridges. Swallows return to their nests year on year, and often have to conduct repair work before they can move back in.

Length
18-20cm (7-9in)

Did you know?

■ Swallows are returning earlier and earlier in recent years, probably as a result of climate change.

■ Before the concept of migration was understood, birdwatchers believed that swallows hibernated under the mud during the winter.

■ Swallows drink water by scooping it from the surface of lakes and ponds in mid-flight. They even sometimes dip right under the water to give their plumage a quick one-second bath.

Swift
Apus apus

Technically speaking, this is one bird that you will probably never find 'in' your garden. Swifts, those aerial acrobats that swoop on scimitar wings across our summer skies, remain airborne throughout their lives, other than when nesting. They feed on the wing, mate on the wing, and even sleep on the wing, climbing to such great heights at night that they've been spotted by aircraft! Summer visitors from Africa, the swift's stay in Britain is among the shortest, arriving in late April and frequently gone before August is out.

Although they never settle in gardens, they can certainly make their presence known.

Nesters in the eaves of taller houses, they shriek around buildings in small groups at astonishing speeds, particularly at dawn and dusk, chasing each other in great excitement. If you've got them near you, then you won't need an alarm clock, as their piercing screams can startle even the soundest sleeper.

Swifts are a good indicator of weather patterns, because they follow the insects upon which they feed. On warm, sunny days they'll be distant specks in the sky, wheeling and turning with mouths agape as they 'vacuum' up anything in their paths. When the atmosphere is heavier and leaden with rain, they'll chase the insects

wn to the height of the treetops, giving you
chance to examine their astonishing acrobatics.
Sometimes, you may find a grounded swift. It
ouldn't be there, though; a swift's legs are too
eak to enable it to take off from the ground, so
oung swift that leaves the nest has only one
ance to get it right. If you find one that seems
be healthy, take it to a first-floor window and
ace it in your palm. Wave your arm gently up
d down so it becomes aware of where it is. It
ould swoop from your palm and fly away, for
at could be up to three years of continuous
ght. If it doesn't, it may be ill, so take it to a vet,
ding it with cat or dog food in the interim.

Facts

Frequency
Modern buildings without accessible eaves have
reduced the swift's nesting opportunities, but
they still arrive in Britain in good numbers.

Identification
Larger than the martins with which they can be
seen, with more scythe-like wings. A sooty-brown
all over, low-flying swifts may reveal their white
throat-patch, more prominent in younger birds.

Song
High-pitched screaming.

Nesting
Urban swifts will readily nest under the eaves
of buildings at least two storeys high. If you fit
a nest box in or by your eaves you may tempt
them to breed there. The nests themselves are
made of whatever the birds can find on the
wing – feathers, straw, seeds and so on – and
cemented together with saliva. In the Far East,
it's the nests of swift species that are used in
the making of bird's nest soup.

Length
16cm (6in)

Did you know?

■ Aerially magnificent though the swift is,
when tired or hungry it can be caught by a
hobby – a small falcon that is so agile it can
even pluck dragonflies out of the air.

■ Swifts pair for life (they have been recorded
as living for more than 20 years). When hunting
for a suitable nesting site, they will frequently
swoop up to the spot and flick it with their
wings, checking for existing
tenants and solidity.

■ A young bird has to know
that it's completely ready to fly
before it leaves the nest, or it'll
crash to the ground and not
be able to take off again. To
test itself, it does 'press-ups'
with its wing-tips. Once
it leaves the nest, it will
set off for Africa within just a
few days.

■ A swift drinks by catching
rainwater as it flies, or by swooping low across
a water surface and scooping up the liquid on
its fly-past.

Tawny Owl
Strix aluco

To attract tawny owls to your garden you really do have to have one of a reasonable size, preferably with a number of mature deciduous trees and backing on to woodland.

Tawny owls, although seldom seen, are the most likely owl to visit your garden, and can be recorded in quite built-up areas, providing there are enough trees around from where it can roost safely during the day. You may catch sight of one during the day, as sometimes roosting tawnies are harried and heckled by smaller birds and forced to move

to another tree. Look for owl pellets at the base of tree trunks, too.

You are more likely to hear the tawny owl, though, as it makes the familiar *hooo* calls, especially in the autumn. They eat rodents, particularly mice and voles, so can be welcome visitors to your garden.

If you have a large garden with a number of mature trees, you could try installing an owl nest box, a long funnel-shaped construction to encourage them to breed. Such a nest box is meant to represent a deep cavity and it should

**'The sluggish, slothful and the
dastarde Owle
About old sepulchres doth daily howle,
And hides him often in an ivy tree,
Lest with small chattering birds he
wrong'd should be'**

From 'Love's Martyr' by Robert Chester

...ed high up on the edge of a stand of trees, and
...ed at 45 degrees to the horizontal, either on
...e trunk or a major branch, and facing east. This
...ll be a heavy nest box and will need at least
...o people to erect it, using ropes and a ladder.
...If a tawny owl box is occupied, it should
...OT be visited in the breeding season! Apart
...m the obvious risk that disturbance may
...use the parents to desert, there is a real
...nger of injury, as a tawny owl will attack any
...ruder and can inflict very serious injuries,
...ecially to the face.

Facts

Frequency

The tawny owl is Britain's commonest and most widespread owl, with about 20,000 breeding in the country each year. It is, however, absent from Ireland and the Isles of Wight and Man.

Identification

Male and female are alike (although she is slightly bigger than he is), with light chocolate mottled plumage streaked with a darker brown. Underparts are lighter but also streaked. The famous owl facial disc is prominent, and the bird frequently sits in a more hunched position than other owls, particularly if it is disturbed during the day.

Song

The famous 'tuwit tuwoo' call is actually the sound of two owls together. The former represents the call of the female, a sharp *kewick*, while the latter is the sound of the male's longer and drawn out hoots. Owlets call for food with a hissing *sheeee*.

Nesting

Although buildings or old crow nests are sometimes used, by far the favourite nesting site is a tree hole, which the owl uses unlined.

Length

37-39cm (14.5-15.5in)

Did you know?

■ As low-flying birds, tawny owls are often hit by cars or their aerials in dark country roads.

Tree Sparrow
Passer montanus

The house sparrow has hit the headlines in recent years for its dramatic population falls, but it's the bird's country cousin that has truly suffered. With a 95% collapse in under three decades (see right), the tree sparrow is truly a bird that is hanging on.

It really is a shocking decline. In the 1960s, the bird was doing quite nicely thank you very much, and *British Birds* magazine recorded that there had been 'a general increase in tree sparrow numbers since about 1958, with an explosive range during 1961-1966 when breeding colonies were established in many parts of the country from which the species had been absent for up to a quarter of a century.

Then came the collapse. Yet all is not lost. Rural gardens in many parts of the UK, particularly the East Midlands and East Anglia can play a part in helping this neat little bird recover at least part of its former status. The tree sparrow is a social bird, gathering in sizeable numbers at the edges of fields, but it will come into gardens – a seed-eater, it will feed from a bird table.

If you have a flock near you – perhaps you li within easy reach of an orchard – you could a try putting out several nest boxes, which mig help the bird expand its local range slightly. T sparrow is a natural hole-nester, and has bee known to take over old sand martin nest sites

Facts

Frequency
There are only about 100,000 pairs left of the tree sparrow, which, like many other birds that rely on traditional farming practices, has seen its numbers plummet in recent years.

Identification
This is a distinctive bird, with a chestnut hood, white cheeks and collar, and a black cheek spot and bib. The wings and the back of the bird are streaked brown with white wingbars. Take a good look at its tail, too – tree sparrows normally hold it in a cocked position.

Song
It chirps, but when flying, makes a *tik-tik* noise.

Nesting:
Tree Sparrows will use nest boxes. Both sexes make the nest out of twigs and leaves, lining it with down, moss and hair. The nest is either a dome or cup-shape built in a hole in a tree, cliff or building.

Length
13-14cm (5in)

Tree sparrows have model relationships, as they mate for life.

Did you know?

■ Tree sparrows are the rural counterpart of the house sparrow, however they are sometimes seen together, especially in the vicinity of untidy farms.

■ There may be a slight increase in numbers in the winter as continental birds occasionally winter in Britain.

■ Outside the breeding season, they will form flocks and sometimes feeds alongside buntings and finches in fields.

Pete Mella, Sheffield Wildlife Trust, advises:

■ The tree sparrow, once a common sight in rural areas, is on the red conservation list for the most threatened of British species. It has suffered a rapid and dramatic drop in numbers, with a staggering 95% decline between 1970 and 1998. The causes of this are thought to be the intensification of specialisation of farming, and increased use of insecticides.

■ Tree sparrows are easy to overlook, as they have very similar voices and habits to their more common, urban cousins the house sparrow, although they are shyer and more active. They can be occasionally be found mixed with flocks of house sparrows, but in Britain are not associated with man.

■ They are widely, but patchily, distributed throughout the UK, but rare or absent in Wales, south-east England and north-west Scotland. Gardens near farmland, or other rural areas, may be lucky enough to get visits from tree sparrows, but hedgerows and woodland edges are much more likely habitats to see them. Like house sparrows, they feed on insects and seeds, and can be seen all year round.

Treecreeper
Certhia familiaris

The treecreeper is one of those birds you only really catch sight of out of the corner of your eye. It's hard to see because its mottled brown back is very well camouflaged against rough bark. True to its name, it's usually spotted shuffling jerkily up tree trunks, clamped to the bark by the sharp claws on the end of its long, strong toes, its body hunched into a gentle outward curve from beak to tail. A glint of its white underparts as it shifts position is most likely to give it away.

Your best chance of spying on treecreepers comes in spring, before the trees come into leaf. They're most active and vocal then too, while establishing breeding territories and wooing mates. It's a joy to watch a treecreeper beavering away, investigating every nook and cranny in the bark for insects and spiders. It uses its fine, down-curved beak like a toothpick to prise beetles, earwigs, aphids, ants, spiders, mites, centipedes and woodlice from their niches, relying on its long, stiffened tail feathers for support as it leans back to gain extra leverage.

As tiny birds, treecreepers are particularly susceptible to the cold. Sometimes, they seek cover in thick ivy or evergreen creeper

Did you know?

■ The treecreeper is the only songbird resident in the British Isles that has a crescent-shaped beak.

■ The treecreeper's white underparts are thought to reflect a little extra light into splits in the bark where its prey is hiding, making it easier for the voracious hunter to find.

■ You need to watch out if you happen to be in the vicinity when a family of fledgling treecreepers leaves the nest for the first time. They have been known to land on the first upright thing they see, be it a tree trunk or person, and begin clambering up.

Facts

Frequency

Broadly distributed in woodland areas, but very easily missed.

Identification

The upper parts are streaked with brown, buff and white, underparts are white; wings are brown with dull yellow and white bars; the juvenile has a shorter, straighter beak than an adult and looks paler, with yellower white areas, faint brown spots on its throat and a dusting of white speckles on its back.

Song

The male is especially vocal from mid February to mid May, in the run up to breeding, singing a high-pitched, tinkling song – *tsee-tsee-tsi-tsi-si-si-si-si-sisisisisisi-tsee* – which gains volume and speed on a declining scale and finishes with a trilling flourish like a miniature bugle call. The sharp *tseee-tseee* contact call it rattles off may betray its whereabouts but often goes unnoticed, as the notes are pitched above the upper range of normal adult human hearing.

Nesting

One of the best hidden nests of all, generally tucked away behind loose bark or in a crevice on a rotten tree trunk; both the male and female stuff the crack with dry grass, moss, lichen, strips of bark and wood chips to form a cup-shaped nest which the female lines with feathers, hair, wool, lichen and spider's silk.

Length

The head and body are no bigger than a wren's, but the long down-curved beak and forked tail extend its overall length to 12.5cm (5in).

Conservation

Harsh winters are a nightmare for a treecreeper – when icy tree trunks are too slippery for it to get a good grip on the bark, and crevices are filled with ice, insect-hunting is impossible. Even under such dire conditions, it rarely comes to bird tables. The best you can do is rub suet and crushed nuts into the rough bark on a tree trunk and hope any treecreepers in the area find it.

'The protectively coloured weevil, the crouching spider clinging to its web, the coccoon enshrouded pupa do not elude its keen eye'

About the treecreeper, from *Birds of the Wayside and Woodland* by Thomas Alfred Coward

rowing on houses and up trees, or in garden utbuildings, but they generally roost behind ose bark, in hollow trees, or in cavities they xcavate in rotten stumps. Regular visitors to favoured tree scratch out a number of holes ound the trunk so that, no matter which way ie wind or rain is blowing, they can always id a dry, sheltered crevice to squeeze into, ittening themselves against the trunk and iffing out their feathers to improve their sulation and enhance their camouflage. In ezing weather, over a dozen of them may iddle together at the same roosting site.

In about 1924, it was noticed in Ireland that ecreepers were showing a preference for osting in large cracks in the thick, fibrous rk of the majestic *Wellingtonia*, a giant lwood pine tree from North America. During e Victorian period, these magnificent trees re widely planted as ornamental features parks and country estates. In the wild, soft, stringy bark serves as a fire blanket, itecting the tree from forest fires. Now, all er Britain, treecreepers are using their beaks d claws to hollow out well-insulated sleeping ommodation in the spongy bark.

Bird nests are incredible feats of engineering. A huge amount of time and effort goes into the choice of site and construction. On these pages, Wildlife Trust and RHS experts explain how you can provide the best possible natural environment for nests and how to give birds a bit of man-made help with nest boxes.

Mary Porter, formerly of Lincolnshire Wildlife Trust, advises:

■ Bird nests have to be built somewhere safe, secret and weatherproof. Do not go looking for nests. Once their 'cover is blown', the birds might be put off or even abandon their eggs or young.

■ Early nesting birds, such as blackbirds, need a snug hedge out of cold weather. Leyland cypress hedging is often a magnet for nesting birds as it provides dense and cosy foliage. However, it is rarely kept thick enough to be safe. As a general rule, if you can put your fist in your hedge, a cat, sparrowhawk or magpie could get in easily, too. A dense, prickly hedge, perhaps containing holly, hawthorn and blackthorn, is ideal.

■ As a brief guide, finches make a cup of twigs, roots and mosses, lined with feathers, woolly seeds or hair, and are

2m (7ft) or more above ground level. Thrushes and blackbirds make a cup constructed from twigs, leaves and mud and can be near the ground or up to 2m above it. A small dome-shaped nest made of mosses and lichens with a lining of feathers, from ground level up to about 2m, and with a round side entrance, is the home of the wren. The male makes several nests to show off to the female, and it is only the one that she chooses that gets the cosy lining. Robins like a cavity, such as a hole in a wall and will take readily to boxes, as do, of course, blue tits, house sparrows, tree sparrows and starlings.

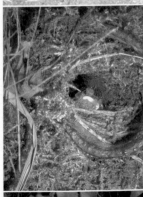

■ Bird nests are protected by law. It is an offence to knowingly destroy one that is being used or built. If you really want to look at a nest, wait until the bare branches of winter reveal it.

What's in your garden?

According to an RSPB survey in 2007, the top 10 commonest garden nesting birds are:

5. Chaffinch

With an average of 1.9 per garden, the chaffinch held its previous position, but numbers were slightly down on last year. Since 1979, numbers have fallen by 36%.

1. House sparrow

The findings revealed that there were an average of 4.4 sparrows per garden. Although at the top of the table, their numbers have fallen by 56% since 1979.

2. Starling

An average of 3.67 starlings per garden and a decline of 76% since 1979. Starlings are noisy characters, usually seen foraging in small flocks. This winter, fewer migrant starlings joined our resident birds from northern Europe because of the milder weather.

3. Blue tit

Blue tits continue to do well, with an average of 2.82 per garden (up from 2.7 last year) and an increase of 16% since 1979. An agile bird, the blue tit is most often seen flitting onto bird feeders. In winter, family flocks are joined by great tits, long-tailed tits and other woodland species, as they search for food.

4. Blackbird

With an average of 2.26 per garden, the blackbird dropped one place from last year. Above-average temperatures during winter meant that fewer blackbirds migrated to the UK from northern Europe, and this may explain the drop in numbers.

6. Collared dove

An average of 1.56 per garden and an increase of 456% since 1979. Originally from southern Asia, collared doves spread from there to the extreme southeast of Europe. They were first recorded in Britain in 1953 and are now a common garden bird throughout Britain and Ireland.

7. Woodpigeon

An average of 1.53 per garden shows that the woodpigeon continues its success in the UK, moving up one place since last year, with an increase of 666% since 1979. They are the largest members of the pigeon family. In flight they have distinctive white wing patches and the tail has a dark band at the end.

8. Great tit

With an average of 1.37 per garden, the great tit saw a slight drop in 2007, although numbers have increased 52% since 1979. Perhaps mild weather has meant that there are better natural food sources available, keeping them away from garden feeders.

9. Robin

An average of 1.26 per garden and a decline of 37% since 1979. Robins are unusual in that they sing throughout the cold season, with both males and females holding winter territories. In winter, some robins migrate from Europe to the UK, but mild European temperatures could explain why fewer were seen this year.

10. Greenfinch

The biggest mover in 2007, the average garden recorded 1.2 greenfinches, compared to 1.7 the previous year. This species is still up 21% since 1979. It's possible that milder weather resulted in better natural food sources and kept them away from garden feeders.

Shelter and nest sites

Birds need sheltered places where they can roost at night or shelter from inclement weather. While some garden birds, such as collared doves and starlings, normally roost in trees, many of the smaller birds prefer the shelter provided by shrubs and hedges, especially those with a dense branch structure. Conifers and evergreen shrubs will give protection against cold winds in the winter. These plants will also provide nest sites for many birds, as will some of the more vigorous climbing plants, such as clematis, ivy and honeysuckle. In order to avoid disturbing nesting birds, pruning and hedge cutting should, where feasible, be delayed until late summer, when young birds will have fledged. Hedgerows and shrubs bearing berries should not be cut back until late winter or after the berries have been eaten.

Nest boxes

Some birds can be encouraged to nest in gardens by providing them with nest boxes. Boxes with an entry hole 3cm (1in) in diameter will be used by tits, while a larger hole of 5cm (2in) will give access to house sparrows and starlings. Open-sided boxes will be used by robins, spotted flycatchers, pied wagtails and wrens. Objects such as old kettles or flowerpots pushed sideways into hedges or dense shrubs may also be used by these birds.

Nest boxes should be sited away from disturbance by humans and cats, and also from bird feeders, so the nesting birds do not have to defend their territory from other birds seeking food. Avoid sites that expose the box to full sun during the middle of the day.

During winter, remove the nest material from used boxes to evict most of the over-wintering fleas, other blood-sucking insects and mites that would make life miserable for the next occupants.

 Helen Bostock of the RHS advises:

Royal Horticultural Society

■ It is a great thing to put up nest boxes for birds, particularly if there is a shortage of natural nest sites, such as old trees with cavities. However, spare a thought for the tree when hanging the box. The traditional method of banging in a nail can be bad news for a tree, creating a site of disease infection. Trees also have a habit of growing around nails. For a more permanent and less intrusive solution, suspend the box from an S-shaped hook. The loop at one end needs to be large enough to slip over a horizonta branch, but at the other only sufficient to hang the box. Plastic-coated steel is durable but non-harmful to the tree, as are specially designed tree hoo that wrap all the way around the trunk.

![The Wildlife Trusts logo]

David North,
Norfolk Wildlife Trust, advises:

■ Whatever did birds do before we've provided nest boxes! No wildlife garden is complete without at least one nest box. They can even be attached to buildings, where they may be used by birds such as house sparrows, starlings or blue tits. The house sparrow and starling are now on the Red List of birds of highest conservation concern, and providing safe nesting sites can really make a contribution to helping species in decline.

■ The easiest birds to attract to nest boxes are the hole-nesting species, such as blue and great tits. There are many good designs available, but why not try making your own? Use wood at least 15mm thick (thick wood provides better insulation and helps the box last longer). Make sure your design includes an easy way to clean out the nest box at the end of the year (always a good idea, to help reduce parasite infection) – a hinged lid is the commonest way of doing this.

Early winter is the best time to put up nest boxes. Birds like blue tits and wrens may use the box for roosting in cold weather and many species select nest sites early in the year. If a box is not used after two years, try moving it to a different location.

■ The range of birds that can be attracted is huge. There are specialist designs suitable for species ranging from tawny owls and barn owls to swifts and tree creepers. To decide which boxes are worth trying in your wildlife garden, assess the habitat and check out what species are present in your area. There is little point putting up nest boxes unless the local habitat provides good feeding areas for young to be raised successfully

■ Be creative! An old kettle in a hedge with the spout facing downwards can make a perfect nest site for robins. However, if you do use unconventional items ensure they are safe and can't, for example, fill with water.

■ Providing nest boxes is one of the simplest ways of attracting birds to breed in your garden and in return the birds will give you many hours of pleasure watching their activities.

Size matters

Many garden and woodland birds nest in holes and may be attracted to a small-hole nest box. The particular species attracted will depend on its local distribution and population, and on the size of hole provided in the nest box. An entrance hole of 28mm in diameter will admit blue tits, great tits, coal tits, tree sparrows and pied flycatchers. A slightly larger hole of 32mm in diameter will also attract house sparrows, nuthatches and lesser spotted woodpeckers. On the other hand, a variety of species may be attracted to an open-fronted nest box placed in a garden, the commonest of which will be robins and wrens, although it could also be used by pied wagtails, spotted flycatchers and black redstarts. If you are considering building a nest box of your own, here are two pointers to bear in mind:

1. If your nest box has an entrance hole, the roof should be hinged for ease of cleaning out, either by a non-ferrous hinge and screws or by a rubber strip. If you have an open-fronted nest box, a hinged roof is not necessary.

2. Small drainage holes should be drilled in the floor.

Waxwing
Bombycilla garrulus

Look at a waxwing in close-up, and it's hard to believe that when in flight, or congregating in a distant tree, it's often mistaken for a starling. These elegant birds, with their pinkish plumage and crest, are impossible to mistake when looked at properly.

The only thing is, you don't often get the chance to do so. During occasional winters, between October and March, sizeable flocks will arrive from Scandinavia and settle in parts of eastern Ireland and Britain, but they pick their sites very carefully, moving steadily westward as they exhaust their supplies of food.

That food is fruit. If you have berrying plants in your garden, such as rowan or hawthorn, then there's always a possibility that a flock of waxwings might drop in one winter. They're quite at home in urban areas, too, so can turn up anywhere.

If you don't have berries, don't despair. Should you hear of a flock nearby, simply hang out some apples from fruitless trees, and you may entice them in. They're fairly confident birds, so you've got a good chance of getting a decent look at this beautiful winter visitor.

The number of visiting waxwings can reach thousands, as it did in the winter of 2004/5 when the berry crop failed in Scandinavia.

Facts

Frequency
As winter visitors to Britain, waxwings rarely breed in Britain, and the number that visits each year varies widely, depending on the food available to them in their breeding lands of north Europe and Russia. When they do arrive, they are most prominent in the east of the country, only moving westwards when food supplies have dried up.

Identification
The waxwing is a highly distinctive bird – plump with a pinkish-buff plumage, palest on its belly. But, it is its crest that makes it so distinctive, along with its black bib and mask around the eye. The tail has a yellow tip and the wings have white and yellow margins with some red, wax-like projections or 'fingers' on its secondary wings, noticeable only on the adult bird.

Song
A high-pitched trilling *srrrrr* call.

Nesting
You will not see the nests in Britain, but both sexes build a cup-shaped nest made of moss grass and twigs.

Length
18cm (7in)

Did you know?

■ Waxwings are so called because of the red tip to their secondary wing feathers, which looks like wax. The function of these tips is unknown, but it could be used to attract a mate. The theory is that the berries that the birds eat can actually deepen the redness, so a brightly coloured wingtip indicates an individual that knows how to find food.

■ A flock might stay around a berry-laden bush for several days.

Willow Tit
Parus montanus

It's bigger than a blue tit but smaller than a great tit. It doesn't have the white nape of the coal tit or the long tail of the long-tailed tit, and its range doesn't overlap with the crested tit. Compared to those members of the family, the willow tit is an easy bird to identify. Unfortunately, there's one other family member, and that's where the confusion begins.

Willow tits were only separated out as an individual species in 1897. Previously they and marsh tits were thought to be the same bird. The main identification points can be found in the 'Facts' box on the right, but the best way to separate them is to look at their black crowns. The willow's is matt, whereas the marsh's is glossy. To help you remember which is which, just think of a woven 'willow mat'.

Yet it would be unfair to think of willow tits solely in terms of their near identical twins. These little birds, which have been suffering a population decline in recent years due to loss of habitat, are engaging creatures, although difficult to see. A denizen of damper areas, such as willow thickets, the edges of lowland peat bogs and around gravel pits, it may take some tracking down, although its nasal, rasping call will help locate it.

f you see one on your bird table, it's almost definitely a marsh tit, as willows rarely come nto gardens. However, if you live near one of their habitats, you may see one on your berry rees in harsh winters. If you're extremely lucky, 'ou may even find that a pair has colonised your est box – to encourage this, fill the box with wood chippings. Willow tits like to dig their way into their nests, and the chippings will give them the opportunity to display this behaviour.

The Mikado, **which has immortalised the bird's name (see bottom right) was actually first performed 12 years before the marsh and willow tit were separated into two species.**

Facts

Frequency
There are 25,000 breeding pairs resident in the UK. It is however listed as a species of red conservation concern as there is less shrubland habitat that the willow tit depends on, leaving it vulnerable to competition from bigger tit species for nesting space. They are mainly found in England and Wales although some are seen in southern Scotland. Ireland has none.

Identification
This species is very similar to the marsh tit, and it was only at the end of the 19th century that they were recognised as separate species. The main identification point is the black cap: in the willow tit it has a matt finish, whereas the marsh tit's is glossy and shiny. The bib is also larger than that of a marsh tit, and somewhat untidier. The bird has white cheeks, a grey-brown back and grey underparts. It has a short bill and bluish legs. Its neck is thicker than a marsh tit's, giving it a more bullish appearance.

Song
A distinctive and nasal *si-si tchay-thcay-tchay* call. The song is musical and warbling.

Nesting
The female excavates a cavity in the tree bark, which she then fills with wood fibres, hairs and feathers.

Length
12-13cm (4.5-5in)

Did you know?

■ The willow tit differs from other tits in its unique ability to carve out a home for itself in dead timber, which is why it has such a thick neck.

■ Despite its similarity to a marsh tit, the two species are rarely seen side by side as they live in different habitats. Willow tits prefer damper wooded areas where there is plenty of decaying wood such as waterside willow scrub. It can also be found in conifer plantations.

'On a tree by a river a little tom-tit Sang "Willow, titwillow, titwillow" And I said to him, "Dicky-bird, why do you sit Singing 'Willow, titwillow, titwillow'"'
From *The Mikado* by WS Gilbert and A Sullivan

Willow Warbler
Phylloscopus trochilus

Willow warblers, it may come as a surprise, are one of the 10 commonest British breeding birds. Around two million pairs arrive in late March and early April from West Africa where they overwinter, and immediately proceed to hide themselves among the woods and copses of the country. There's a good chance that you've had them in your garden at some time – it's just that you may easily have overlooked them.

Alternatively, you might have thought that they were chiffchaffs. The two species are so similar that even experienced birdwatchers are uncertain which they've seen, unless they get a really good look. Size, colour and markings are virtually identical at a distance, although closer up the willow warbler is a slightly lighter-coloured bird, with pale brownish legs compared to the chiffchaff's generally darker legs, and a slightly more prominent eye-stripe.

Not much to go on visually, but fortunately when the two birds open their mouths to sing during the spring, all doubt falls away. Where the chiffchaff simply sings its name, the willlow

> If a willow warbler is disturbed on the nest, she will fly away and pretend to be injured in an attempt to lure away the predator.

Facts

Frequency
Britain's commonest summer migrant, the willow warbler is found throughout the British Isles.

Identification
An active little warbler, it is greenish brown above with a hint of yellow on its breast. The cream stripe above the eye stands out more than in other leaf warblers, while the legs and feet are paler than those of the chiffchaff.

Song
A lyrical cadence of falling notes marks the willow warbler out from the chiffchaff. Calls include a thin *whoo-eet*, and the more alarmed *pew pew*.

Nesting
Unusually for a bird of this size, the willow warbler sometimes builds its grassy, mossy, leafy ball-shaped nest very close to the ground. At other times it nests in hedgerows.

Length
11cm (4.5in)

'A willow wren still remembered his love, and whispered about it... so gentle, o low, so tender a song that it could carce be known as the voice of a bird.'

om *The Open Air* by Richard Jeffries ('willow wren' is e name once given to this warbler)

arbler lets forth a clear trill that slides down the ale, the perfect musical clue to its identity.

Spring is not just the best time to hear willow arblers, it's the best time to see them. Like the oldcrest, it sometimes joins flocks of foraging tits, ut it generally works alone, flitting through the aves and twigs of hedge and tree, plucking insects om the undersides of foliage and sometimes overing to snatch an insect from mid-air. During his time they're most likely to stay in their ountryside haunts, but as autumn approaches and ey need to stock up on fuel for the long journey ack to Africa, they're more likely to visit gardens r the juicy greenfly they can find there.

Did you know?

■ To help fuel itself for its long flight back to Africa, the willow warbler will sometimes supplement its insectivorous diet with soft fruit and berries.

■ Droughts in the bird's wintering grounds of sub-Saharan Africa have led to a recent decline in numbers.

■ The willow warbler is one of the cuckoo's hosts, the tiny bird eventually rearing a chick several times its own size.

■ It was Gilbert White, the famous 18th-century naturalist from Selborne, who first realised that the chiffchaff and the willow warbler were different species.

Woodpigeon
Columba palumbus

There's something of the old-fashioned military about the woodpigeon. Big, bluff and bloated, it struts around your garden like Colonel Blimp, an official stripe on each wing and its chest pumped out as if it owned the place. Yet the slightest disturbance and it clatters into the air with plenty of bluster and in apparent confusion, like an old dog who has not learnt the new tricks of the trade. Once in the air,

it transfers from army to air force. There's a wing-thumping climb up to the sky, followed b a steady swoop with wings outstretched in a V shape, resembles an old fighter plane at an aeri show, putting itself through its paces one last time as its engine struggles to keep it aloft.

Bumbling it may appear to be, but there's mor to the woodpigeon than meets the eye. This, aft all, is a bird that has spent generations upon

> **'And the deep mellow crush
> of the woodpigeon's note
> Made music that sweeten'd the calm.'**
>
> From 'Field Flowers' by Thomas Campbell

generations being shot at, and it's had to build in a number of safety measures to survive. The woodpigeon that clatters its way through the twigs of a tree as you approach appears to have been caught by surprise, but in fact it's merely trying to turn the tables. It's been keeping you in its eye all along, and the frantic nature of its escape is designed to startle you, throwing you off your guard, and giving the bird the extra time that it needs to get its great bulk into the air and escape. It also serves as an alarm to other birds.

The woodpigeon can rarely be described as the gardener's friend, however. Vegetable plots are particularly vulnerable to its voracious appetite, so one solution is to provide a distraction. Seed scattered nearby is much easier for the woodpigeon to handle, so a small pile should keep the bird away from your valuable veg. If you'd rather keep the birds away completely, though, a wind-chime can often work well on this nervous bird.

If you do get woodpigeons coming in, and you have a pond, watch to see how they drink. Most birds scoop mouthfuls of water up in their beak then tip their heads back to swallow it, but woodpigeons are actually capable of sucking the liquid up through their beaks, as if through a straw. The bird needs plenty of water to drink, as it does not get much moisture from the seeds and grains that it feeds upon. In fact, the woodpigeon actually has to moisten its food before it eats it. It has a special crop in its gullet into which it swallows its food, where it is moistened before entering the stomach. The crop also produces a special milk during the breeding season which the pigeon uses to feed its young.

Facts

Frequency
A very common bird, the woodpigeon numbers some five million birds at the end of each breeding season. It can be found across Britain and Ireland wherever there are woody areas.

Identification
This is the biggest of Britain's pigeons. It has a pencil-grey head and a white collar atop a stout body with a dark lavender breast. In flight, its white wing-bars are very conspicuous.

Song
A five note cooing, sometimes anthropomorphised as 'take two toes, Taffy', makes up the song which can be heard as early as February.

Nesting
The woodpigeon's nest is surprisingly flimsy for such a bulky bird. A loose collection of twigs is meshed together at the top of a tree or hedge, and that's about it.

Length
41cm (16in)

Did you know?

■ Woodpigeons are fairly successful breeders, and the spring-time population can easily have trebled by the end of the summer.

■ A young pigeon is called a squab.

■ The barrel-chested appearance of the woodpigeon is due to the bird's powerful chest muscles which it needs to be able to take off and raise its bulk. It weighs an impressive half a kilogram or more.

The woodpigeon's crop is astonishingly voluminous. It can hold up to 1,000 grains of barley, or alternatively, 60 acorns.

Wren
Troglodytes troglodytes

An interesting Latin name is the wren's. *Troglodyte* means cave-dweller, and this little bird that tucks itself into the ivy or roosts in tiny wall crevices certainly lives up to that concept. Yet cave-dwellers tend to keep themselves to themselves, and in this respect, wrens are completely different. The next time you hear a loud song, with rapid notes that leap up and down the scale and are punctuated by machine-gun-like rattles, you might think it comes from a large bird. Not so. The wren belts out its militar[y] tune with astonishing gusto and force for such little lungs, and, like the Wizard of Oz, comes across as being much larger than it actually is.

Wrens are common birds, with over eight million pairs breeding each year. Yet they're vulnerable, too. A harsh winter, such as that in 1962/3, can destroy populations, as the little bird has no resistance to extreme cold, other than to bunch up tight in nest boxes trying to keep themselves warm. Astonishingly, the record number of wrens roosting in one nest box is 63!

Its size does provide it with one advantage, though, enabling it to reach parts of your garden that other birds can only dream of. Slipping between twigs and bramble, it raids crevices, nooks and crannies for flies, spiders, woodlice, aphids and a host of other invertebrates that might otherwise consider themselves safe. The wren is truly a gardener's friend.

Talking about safety, young wrens tend not to understand this basic principle. Once they leave the nest, they're difficult to round up again for the protection of the night-time roost, and you may often hear adults *chik*-ing at their young to get back into the nest late into the evening.

Wrens were one of the great exploiters of bomb sites after World War II, enjoying the plant growth and lack of people. Once the sites were developed, the wrens moved on again.

Facts

Frequency
With eight million breeding pairs across the entire country, islands included, the wren is one of our commonest and most widespread birds.

Identification
Absolutely tiny, with a mottled brown plumage, no neck, and cocked tail. Flight tends to be in short bursts low across the ground.

Song
A rattling beat at surprising strength, invariably with a fast-paced staccato passage included. Alarm calls are hard *chik*s and *chur*s.

Nesting
The male builds several well-hidden ball-shaped nests from leaves, grass and moss, then lets his mate pick her favourite. She then lines her chosen home with feathers and hair.

Length
9-10cm (3.5-4in)

Did you know?

■ A wren is so light and buoyant that if it slips into water it can swim its way out again.

■ According to folklore, when the birds held a contest to see who could fly the highest, the eagle appeared to have won, until a tiny wren emerged from his feathers and flew just a little bit higher.

'Among the dwellings framed by birds
In field or forest with nice care
Is none that with the little Wren's
In snugness may compare'

From 'A Wren's Nest' by William Wordsworth

Companion

Hedges

RHS' Tony Dickerson advises:

Royal Horticultural Society

Hedges are wonderful ways of delineating borders in your garden, or between you and your neighbour's garden, and they provide a wealth of opportunities for wildlife to thrive as well.

The RHS provides a very useful advisory leaflet on hedge preparation and planting, as well as many other leaflets on all types of subjects. You can find them by visiting www.rhs.org.uk.

The ideal time to plant a deciduous shrub, whether bare-rooted, root-balled or container-grown, is from autumn to early winter. Evergreens also establish well if planted in autumn or early spring. Do not plant in frozen or waterlogged soil and avoid the summer months when plants are likely to dry out.

Ground preparation

Dig over an area 60-90cm (24-35.5in) wide and one spade-blade deep. Do not double-dig because the disturbed soil will resettle, resulting in the hedging being too deep once planted. Instead, lightly fork over the base and sides of the trench to break through smeared surfaces and aid drainage. Remove all weeds. For long runs of hedging, a rotovator may be needed. This should be done while the soil is relatively dry in early autumn, especially on heavy clay soils, otherwise the rotors are likely to smear the soil beneath the level of cultivation, impeding winter drainage.

Do not add organic matter to the bottom of the trench as it decomposes, causing the shrub to sink. On fertile soils it is usually unnecessary to add additional organic matter at all and may be counter productive as it discourages the roots from growing out into the surrounding soil. However, on sandy or heavy clay soils, organic material such as garden compost, or a proprietary tree and shrub planting mix, can be incorporated into the backfill (the soil dug out from the hole) to improve soil structure, or spread over the soil surface and lightly forked or rotovated in. It is not beneficial to apply any fertiliser until the following growing season.

On poorly drained soils, fork horticultural grit and composted bark into the bottom of the trench and form a ridge about 15cm (6in) high to plant into. Do not add organic matter or grit to the bottom of a trench, as this merely creates a sump. Soils that become waterlogged in winter may require a permanent drainage system.

Call and Song identification

The art of bird call identification is seen by many as a major landmark in developing as a birder. However, if you ever achieve the heady heights of being referred to by your friends as someone who is: "good with bird call", just exactly what does that mean, and how did you get there?

When you start to learn bird sound many parallels can be drawn with learning a language, though unfortunately not with the obvious, for instance listening out for a particular word, the meaning of which, gives a clue to the context. Rather the parallels are with the less tangible elements of sound: tone, rhythm and intonation.

As such, the act of learning bird call can seem a daunting task. However, the first step is your resolution to make bird call a priority. Let me try to make to it a bit easier from a beginners perspective:

● Firstly, remember to focus on what you can identify already; there are loads of birds whose calls in fact we know, but haven't labelled yet. Begin by labelling these! By this I mean verify beyond all doubt, that you are hearing a 'x', and don't satisfying your self with statements like, 'I think it's probably a…'. Though this may sound trivial it is the first step to building the confidence to progress.

● Secondly, start to learn the calls of the commonest species first. Often the art of identifying by sound has a substantial element of elimination. You are learning to filter out common species to allow you to focus on those a little more unusual. In tandem the other area to focus on in this 'filter' phase are the really extreme calls. Even if not particularly common they are easy to learn.

● Thirdly as you improve, remember it is in identifying the differences between similar sounding species, that is the real skill. The thing to remember here, is that it's not just about learning the sounds in isolation. The way to learn the difference between two sounds is to compare them. By actively comparing and contrasting similar sounding species in the learning process, the subtle differences that one would not pick up by listening to the call in isolation become much more apparent.

At the end of the day before you can get anywhere with call identification, you need to have some patience. The great thing is that in today's world with the array of technology about, help is close at hand. CDs make a good start, but there is a fundamental flaw in their ability to enable you to select the bird you want to listen to quickly. By far the best product that the authors of this book have come across is a really clever device called BirdVoice, that instantly lets you compare two species. By pointing the magic BirdVoicePEN at the name of the bird you instantly hear its calls and song. By listening to them repeatedly you will quickly pick up the subtle differences.

You can have a look at BirdVoice at www.birdvoice.co.uk
or contact the company directly on 0845 600 1361.

Pheasant
Phasianus colchicus

Abundant though pheasants are, they rarely come into gardens, even those that back onto fields. As they're in danger for several weeks of the year of being shot at, this is perhaps unsurprising. Nonetheless, the occasional bird will stray in, particularly during colder winters, and add a touch of oriental charm to your garden. Pheasants may have been here for centuries, but they were originally introduced, as game birds, from China.

Pied flycatcher
Ficedula hypoleuca

As this is a bird of mature woodland, the only real chance of spotting it in your garden is during its spring and autumn migratory passage en route to the west country. A busy little bird, about the size of a sparrow, and with a striking black and white plumage in the case of the male, you may see it flitting among trees on the hunt for insects, before setting off on the next leg of its journey. The female (below) is brown.

Redpoll
Carduelis flammea, C. cabaret etc

There was a time when a redpoll was a redpoll was a redpoll. Now it's been split into mealy redpolls, lesser redpolls, and various other races that have made identification very difficult. In a nutshell, though, all redpolls are lively little finches with, as their name suggests, red crowns. Their populations have fallen heavily in recent decades, but indications are that they might recover and become a garden possibility once more. Scottish gardens have the best chance of hosting them.

Turtle dove
Streptopelia turtur

Once a reasonably manageable sight during the summer months, the turtle dove has gone on a serious decline in recent decades. The gentle, purring *turr-turr* that gives it its name is rarely heard these days among the woodland edges and bushes that it once frequented. Nonetheless, several thousand pairs still breed here, and there's still a chance that one might be seen hunting for insects in your garden, should you live near a wood.

Explore
Lee Valley Regional Park

With an incredible eight Sites of Special Scientific Interest and the Lee Valley Special Protection Area and Ramsar site, Lee Valley Regional Park is a key regional wildlife destination serving London, Hertfordshire and Essex. Green spaces, nature reserves, heritage sites and country parks run both sides of the beautiful River Lee, making the 26 mile Regional Park the perfect place for bird watching, fishing, cycling and walking.

Offering visitors a wide range of rich flora and fauna throughout the seasons, there are plenty of opportunities to see wildlife up-close, such as orchids in the summer or rare water birds like the elusive Bittern in the winter. Why not take a visit to the unique WaterWorks Nature Reserve to see its different wetland habitats or experience the stunning scenery and wildlife of River Lee Country Park?

Lee Valley Regional Park also offers excellent recreation opportunities, including golf, horse riding and ice skating - all allowing you to combine sport and nature for a great day out. There is a range of accommodation within the Regional Park, including a top class youth hostel and ideally situated caravan and campsites, providing reasonably priced accommodation for a fantastic weekend away.

For more information about Lee Valley Regional Park and what you can do and see there, call 01992 702 200 or visit www.leevalleypark.org.uk

1967-2007 celebrating 40 years

Lee Valley Park

Open spaces and sporting places

Europe's only dedicated
PARROT ZOO

The National Parrot Sanctuary
Conservation & Education Centre

Come & See Over 1200 Rescued Parrots

THE NATIONAL PARROT SANCTUARY

Cafe / Restaurant • School Visits • Guided Tours
Childrens Parties • Parrot Experience Days • Plus much more

Tel: 01754 820107 / Fax: 01754 820406
www.parrotsanctuary.co.uk
Dickonhill Road, Friskney, Lincolnshire, PE22 8PP
Follow the brown tourist signs from the A52 / A16

Open Daily 10am - 5pm
Adult £4.60 / Child £3.00 / OAP £3.00

"The Most Tropical Place In Lincolnshire"

English Tourism Council

QUALITY ASSURED
VISITOR
ATTRACTION